GEMS OF BEIJING ANCIENT
ARCHITECTURE

北京古代

建筑精粹

北 京 市 文 物 局
《北京古代建筑精粹》编委会 编

北京出版社 出版集团
BEIJING PUBLISHING HOUSE (GROUP)
北京美术摄影出版社

# 编　委　会

**策　　划**

梅宁华　吴雨初

**主　　编**

孔繁峙　钟制宪

**副 主 编**

崔国民　舒小峰　李清霞

**执行副主编**

王玉伟　韩　扬　于福庚

**艺术总监**

左汉桥

**执行编辑**

王有泉　侯兆年　张肇基　程阳阳

**总体设计**

刘金川

# 序

北京历史悠久,又有四代封建王朝在此建立首都,遗留下品类丰富的古代建筑。从帝王施政与生活的宫殿、休憩娱乐的园林到死后埋身的陵墓;从祭天拜祖的皇家坛庙到晨钟暮鼓的宗教寺观;从险峻雄伟的长城到朴素无华的仓廪;从气派的王侯府邸到恬静的普通民居……无一不铭刻着我们灿烂古代文明的印记,印证着北京悠久的历史,传播着我们优秀的传统文化。

目前北京有各级文物保护单位885处,其中的故宫、长城、天坛、颐和园等文化遗产久已享誉世界。为了将包括它们在内的更多的北京优秀古代建筑更加充分地向世人展现,北京市文物局和北京出版社出版集团合作编著了这本《北京古代建筑精粹》。

本书是一部全面、系统地反映北京现存的各类古代建筑精品的书籍。书分上、下两卷,在真实地记录下这些不可再生的古代建筑的基础上,编著者力图通过本书宣传北京,增强人们热爱文物的意识,使人们在欣赏这些古代建筑的同时,也能够感受到历史责任,从而推进文化遗产的保护工作,使它们在漫漫历史长河中释放出更加璀璨的光芒。

北京市文物局
《北京古代建筑精粹》编委会
2007年10月

# Preface

With a long history and the capital of four feudal dynasties, Beijing has abundant ancient architecture. From the palaces where the emperors lived and worked and the gardens where they enjoyed leisure and entertainment to the tombs where they were buried after death, from the royal altars where they worshiped Gods of Heaven and ancestors to the Buddhist and Taoist temples where bell rang in the morning and drum sounded in the evening, from the steepy and grand Great Wall to the simple and unadorned storehouses, from the noble's aristocratic-style mansions to the peaceful common people's dwellings, all of these architecture bears the imprint of our splendid ancient civilization, proving the long history of Beijing and spreading our excellent traditional culture.

In Beijing, there are 885 relics under preservation of various levels, many of which are highly admired and praised by people of all countries, like the Forbidden City, the Great Wall, the Temple of Heaven and the Summer Palace. *Gems of Beijing Ancient Architecture* is compiled by Beijing Municipal Administration of Cultural Heritage and Beijing Publishing House (Group) in order to let more people know excellent ancient architecture of Beijing completely.

As an all-round and systematic reflection of various ancient classic buildings existing in Beijing, the book is divided into two volumes. Based on a true record of the nonrenewable old buildings, it is intended to publicize Beijing, build up people's consciousness of loving cultural relics, make people aware of their historical responsibility when they appreciate these old buildings, promote the protection of cultural heritages and make the heritages more brilliant in the long history.

Beijing Municipal Administration of Cultural Heritage
Compilatory Board of *Gems of Beijing Ancient Architecture*
October, 2007

# 凡例

一、本书全面、系统地介绍和展示北京现存古代建筑当中，具有代表性和典型性的中国传统形式的古代建筑102处（遗址及复建的古代建筑未收录）。共收录古代建筑照片700余幅，线图20余幅。

二、根据使用功能，本书将北京的古代建筑分为9大类：即宫殿、坛庙、园林、学府衙署、府邸宅院、寺观、城垣、陵墓和其他。每类当中又按照历史、规模和代表性等原则有所区分和侧重。其中"塔"因其多数为寺观的一部分，其建筑功能也是服务于宗教，所以归入寺观，不单独分类。

三、本书分为上、下两卷。上卷卷首为总论《北京地区古代城市和建筑发展概况》，入录宫殿、坛庙、园林、学府衙署、府邸宅院5个章节。下卷入录寺观、城垣、陵墓和其他4个章节，卷尾附录"北京市各级别文物保护单位名录"和"北京地区古代建筑修缮工程大事记"。

四、本书在每一类古代建筑内容前都有一篇整体介绍其历史发展和建筑特色的文章。

五、除图片说明外，部分重要建筑物还附有简短的文字说明和建筑结构线图。

# 总目录
## CONTENTS

上卷
VOLUME I

序
Preface

北京地区古代城市和建筑发展概况
An Outline of Ancient City of Beijing
and Its Architecture     14

宫殿
Palaces     32

坛庙
Altars and Temples     98

园林
Gardens     156

学府衙署
Academys and Government Offices     240

府邸宅院
Mansions and Residences     266

下卷
VOLUME II

寺观
Monasteries and Temples     8

城垣
City Defense Installations     184

陵墓
Tombs     226

其他
The Others     272

附录
Appendices     314

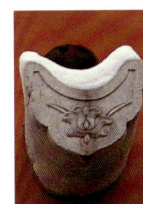

后记
Postscript     341

北京古代建筑精粹

GEMS OF BEIJING ANCIENT
ARCHITECTURE VOLUME I

北京市文物局
《北京古代建筑精粹》编委会 编

北京出版社 出版集团
BEIJING PUBLISHING HOUSE (GROUP)
北京美术摄影出版社

# 上 卷 目 录

北京地区古代城市
和建筑发展概况 14

宫殿 32
北京的宫殿 34
故宫 44

坛庙 98
北京的坛庙 100
天坛 108
地坛 124
日坛 126
月坛 127
社稷坛 128
先农坛 130
凝和庙 136
太庙 138
孔庙 144
历代帝王庙 152

园林 156
北京的园林 158
北海及团城 166
中南海 194
景山 200
颐和园 206

学府衙署 240
北京古代的学府衙署建筑 242
国子监 248
顺天府学 256
金台书院 258
皇史宬 260

升平署戏楼 262
古观象台 264

府邸宅院 266
北京的住宅建筑 268
恭亲王府 274
醇亲王北府(醇亲王府) 290
醇亲王南府(醇王府) 297
庆王府 299
孚王府 300
郑王府 304
克勤郡王府 310
和敬公主府 316
涛贝勒府 318
那王府 322
霱公府 326
东城区张自忠路23号 (孙中山行馆) 332
宣武区珠市口西大街241号(纪晓岚故居) 334
东城区圆恩寺后街7号四合院 336
西城区富国街3号四合院 340
东城区帽儿胡同7号、9号、11号(可园) 342
东城区府学胡同36号、交道口大街136号四合院 344
东城区礼士胡同129号四合院 352
西城区西四北大街六条23号四合院 358
东城区东棉花胡同15号院及拱门砖雕 360
西城区西四北三条11号四合院 364
西城区前公用胡同15号四合院 366
东城区黑芝麻胡同13号四合院 370
东城区鼓楼东大街255号四合院 372

# Contents of Volume I

An Outline of Ancient City of Beijing
and Its Architecture   24

PALACES   32
Palaces in Beijing   40
The Palace Museum   44

ALTARS AND TEMPLES   98
Altars and Temples in Beijing   104
TheTemple of Heaven   108
The Altar of Earth   124
The Altar of the Sun   126
The Altar of the Moon   127
The Alter of Land and Grain   128
The Altar of the God of Agriculture   130
Ninghe Temple   136
The Imperial Ancestral Temple   138
The Temple of Confucius in Beijing   144
The Temple of Successive Emperors   152

GARDENS   156
Gardens in Beijing   162
Beihai and Round City   166
Central and South Seas   194
Jingshan Hill   200
The Summer Palace   206

ACADEMYS AND
GOVERNMENT OFFICES   240
Ancient Academys and Government
Offices Architecture in Beijing   245
Guozijian (the Imperial College)   248
Shuntianfu College   256
Jintai Academy   258
Huangshicheng (the Imperial Archives)   260
Opera Theater in Shengpingshu   262
The Ancient Beijing Observatory   264

MANSIONS AND RESIDENCES   266
Beijing's Residential Architecture   271
Prince Gong's Residence   274
Prince Chun's North Residence   290

Prince Chun's South Residence   297
Prince Qing's Residence   299
Commandary Prince Fu's Residence   300
Prince Zheng's Residence   304
Commandary Prince Keqin's Residence   310
Princess Hejing's Residence   316
Beile Tao's Residence   318
Prince Na's Residence   322
Mr.Yu's Residence   326
The Quadrangle at No. 23 Zhang Zizhong
Street in Dongcheng District   332
The Quadrangle at No.241 Zhushikou Street
in Xuanwu District   334
The Quadrangle at No.7 Yuanensi Backstreet
in Dongcheng District   336
The Quadrangle at No. 3 Fuguo Street
in Xicheng District   340
The Quadrangle at No. 7, No. 9 and No. 11
Maoer Hutong in Dongcheng District   342
The Quadrangles at No.36 Fuxue Hutong
and No.136 Jiaodaokou Street
in Dongcheng District   344
The Quadrangle at No. 129 Lishi Hutong
in Dongcheng District   352
The Quadrangle at No.23 Xisi Beiliutiao
Street in Xicheng District   358
The Quadrangle at No.15 Dongmianhua
Hutong in Dongcheng District   360
The Quadrangle at No.11 Xisi Beisantiao
Street in Xicheng District   364
The Quadrangle at No.15 Qiangongyong
Hutong in Xicheng District   366
The Quadrangle at No.13 Heizhima
Hutong in Dongcheng District   370
The Quadrangle at No.255 Guloudong Street
in Dongcheng District   372

# 北京地区古代城市和建筑发展概况

北京是中华人民共和国的首都，是中国政治、经济、文化的中心。北京历史悠久、文物众多，是世界著名的古都和历史文化名城。早在3000多年前，北京便出现了城市，从房山琉璃河发现的商周古城址，到元代辉煌的大都城和明清北京城；从巍峨的宫殿、神秘的坛庙、优美的园林，到雄伟的长城、严密的城防体系、静穆的皇陵、成片的四合院……北京集中了中国古代建筑最完整、最精华的遗产。现有世界文化遗产6处，全国重点文物保护单位98处，北京市文物保护单位200多处，各区县文物保护单位500多处。

## 一、城市的形成与发展——史前至唐代北京的城市与建筑

远在70万年前，北京就有人类生活，这就是上世纪20年代在北京西南房山区周口店发现的"中国猿人北京种"，俗称"北京人"。此时的人类主要居住在洞穴中，还不能称其为建筑。进入新石器时代，北京地区发现的代表遗址是"东胡林人"的墓葬。"东胡林人"的墓葬位于北京西郊门头沟区东胡林村西侧，清水河谷西岸的黄土台地上。经鉴定，"东胡林人"的生活年代距今约1万年，属于新石器时代早期。由于"东胡林人"的遗骨和文化遗物在台地上发现，说明北京地区的古人类逐渐移居到平原上生活。这个时期的人类遗址还有平谷县的上宅和北埝头遗址、房山区镇江营遗址、昌平雪山遗址等。其中，北埝头遗址位于平谷县城西北7.5公里。这里发现了10座新石器时代的居住遗址，这些房屋为圆形或椭圆形的半地穴式，直径一般在4米左右。经测定北埝头遗址距今为六七千年，是迄今为止北京地区发现最早的人类居住遗址，为研究北京地区人类早期建筑提供了宝贵的资料①。

1984年，考古工作者在北京房山琉璃河乡董家林村，发现了一处规模很大的商周遗址，学术界定名为琉璃河遗址，经研究学者们一致认为这个遗址就是周朝初年燕国的始封地。琉璃河遗址东西长3.5公里，南北宽1.5公里。遗址分为居住址、墓葬区和古城址。古城址东西长850米，城南

部由于大石河泛滥冲成洼地，所以仅余北城墙遗址，南北长度不清。城墙结构有主墙、内附墙和城外平台，主城墙宽约2.6米，内附墙在主墙内侧，紧贴主墙面，城外平台在主城墙外侧，低于主墙，呈平面状。城墙结构为夯土版筑而成。在城墙周围均有护城河围绕。据研究测定，此城墙建成使用是商代后期，在周朝初年即遭废弃。结合地理位置和文献资料，这个古城址即西周前的古代燕国遗址②。

至春秋时代，燕国为北方主要诸侯国，以蓟城为国都。蓟城的地理位置在北京旧城的西南广安门、白云观一带，建国后这一带不断出现带有燕国纹饰的瓦井圈，学者们基本认定这一带是春秋时期燕国的蓟城③。当年由于北方少数民族山戎的威胁，燕国曾一度迁都临易（今河北省雄县）。当时燕国所辖北至燕山山脉，南至大清河，西边包括西山，东到海边。今天北京的延庆县及怀柔县北部山区是山戎部族活动区域。现延庆县玉皇庙等地发现的山戎墓葬群证明了当年山戎部族的活动范围。

战国时期，燕国为战国七雄之一。燕都仍沿用春秋时地处北京西南的蓟城。除蓟城外，燕国还有"中都"和"下都"。燕中都城位置在北京房山县窦店以西，1959年在此处发现一座土城址，呈方形，有内外两层城垣，外城堆积土围，为古城外廓，内城东西长1230米，南北宽1040米，据考证此城是汉代的良乡城或燕中都城。燕下都仅有文字记载，未发现古城墙遗址。战国后期，燕国向北扩展疆土，修筑了燕长城，今河北省宣化、怀来以北，还留有战国时期古长城遗址。

秦朝是中国历史上第一个中央集权制的封建国家。据史书记载，秦朝将蓟城以北的上谷、渔阳等郡内原有的燕长城进行修补和改造，建成北京地区的秦长城。现存北京地区秦长城的遗址仍在调查和考据。

西汉时期地方行政制度为郡国并行制，地方行政管理即设郡，又封立诸侯王国。郡与王国下设县和诸侯国。今北京地区分属广阳国、上谷郡、渔阳郡、涿郡、右北平郡。汉武帝在北京地区设幽州刺史部。公元8年，王莽篡夺西汉政权，郡县名和诸侯国名更改了许多。公元25年，建

立东汉，恢复西汉旧制，改蓟城为幽州。北京在东汉始称幽州。

西汉时，北京地区属燕地蓟城，长期封立王国，财富有较雄厚的积累。从史书记载中，可知当时蓟城建筑有相当规模，蓟城有连接通往京师大道的南门，蓟城四周有带城楼的大门。燕王的宫殿有宫城，宫城内有朝宫，曰万载宫；有朝殿，曰明光殿；宫旁有溪泉和池沼，说明建筑规模已相当宏大。

1974年，在北京丰台区大葆台村发掘出一座西汉燕王(或广阳王)墓葬，该墓平面呈"凸"字形，南北长23米，东西宽18米，深3.7米，有封土、墓道、甬道、内外回廊、黄肠题凑、前室(便房)、后室(椁室)和棺椁等部分组成。在墓道的北部埋葬着马车3辆，马11匹；外回廊埋着3匹马和金钱豹。黄肠题凑由15000根柏木堆积围绕前后墓室一周。在墓中发掘出大量的随葬品，包括陶、铜、铁、玉、玛瑙、漆器、丝织品等，代表了西汉北京地区经济和技术的发展水平。

西汉时期，北京地区建筑已形成规模，出现了一些居民集中的城市。经多年考古发现，在北京郊区有许多汉代古城遗址，如房山区良乡广阳城村的汉代广阳县城遗址、窦店汉代土城遗址、长沟汉代土城遗址、周口店蔡庄古城遗址、卢村古城遗址、昌平区汉代军都城遗址、平谷县北城子村古城遗址。这些古城遗址分布较密，最近的几座城址之间相距不过30里，可见当时经济的繁荣程度。

东汉时期在北京地区遗存的建筑遗迹主要是墓道石阙和墓穴。较为知名的有在北京西郊八宝山发现的东汉和帝永元十七年(105年)的"汉故幽州书佑秦君神道"石阙，发现的17件石阙构件分别刻有铭文和纹饰。东汉的墓葬规模也很宏大，如北京怀柔县城北发现的9座东汉墓中，多室墓3座，复式墓2座，单室墓4座。其中东汉晚期的31号多室墓规模最大，有墓道和甬道，甬道通向中室，中室南端左右各有一耳室，中室左右又各有一棺室。从墓葬的规模可推测出当时建筑规模和发展水平。

汉代的建筑实物今已无存，但从墓室的随葬品中出土了许多陶制房屋模型，应该是墓主人生前所居房屋的式样。在怀柔城北31号墓室中出土的陶楼非常引人注目。陶楼分上下两层，上层正面两旁各有一长方形窗，中间有一长方形菱花窗。正面檐下有两个斗拱，屋顶为庑殿式，瓦垄头为兽面纹，4条垂脊，垂脊头用兽面纹瓦当封护，正脊两端有鸱尾。这说明汉代建筑已具备了中国传统建筑的雏形。

东汉献帝延康元年(220年)十月，曹丕代汉自立，国号魏，曹魏统治了46年，今北京地区仍属幽州。曹魏以后为西晋。西晋实行州、郡、县三级制。今北京地区为西晋幽州所属的燕国、范阳国、上谷郡。

南北朝时期，北京地区属北魏统治，地方行政制度仍为州、郡、县三级制。今北京市境分属幽州管辖的燕郡、渔阳郡、范阳郡、燕州上谷郡、安州密云郡、广阳郡等。

581年，隋朝统一中国。隋文帝改地方行政建制为州县二级制，大业初年又改为郡。北京市境分属于涿郡、安乐郡和渔阳郡，其中蓟城归涿郡管辖。618年，唐朝建立，改郡为州，北京地区仍为幽州管辖，下治蓟城。唐朝时期，蓟城已成为北方地区的军事重镇和贸易中心。

幽州蓟城(今北京)在唐代又称幽州城，是唐代幽州的治所。《太平寰宇记》记载："蓟城南北九里，东西七里，开十门"。据此可知，唐幽州城是一座南北略长、东西略窄，平面为长方形的城池。据学者考证，唐幽州城周围32唐里，约合今24里左右。近年来，考古学者根据历年发现的唐人墓志对幽州城进行了考证，初步认定了唐代幽州城的方位和范围。唐代幽州城的四至大致是：东城墙位置在今北京宣武区烂缦胡同与法源寺之间的南北一线；西城墙在今北京宣武区会城门村以东的小马厂和北京钢厂院东侧和甘石桥东侧南北一线；南城墙在今北京宣武区白纸坊东、西大街一线；北城墙在今北京西城区白云观一带。

根据史书和近年发现的墓志铭记载，当年唐幽州城内有26个坊，每坊有门楼，楼上有坊名的牌匾。"坊"是唐代城市的基层单位。"坊"是由街道分割而成的封闭居民区，坊的格局一般为方形，四边有墙，各开一门，两门之间有巷相通。唐代幽州城居民按坊居住。据近年出土唐墓志铭记载，当年幽州城中有罽宾坊、卢龙坊、肃慎坊、花严坊、辽西坊、铜马坊等24坊的坊名，与文献所载26坊坊名相差无几。

唐朝时期，佛教兴盛，在幽州地区寺庙数量不断增加，比较著名的有马鞍山慧聚寺(后称戒坛万寿寺)、白带山云居寺(今房山云居寺)、淤泥寺(后称鹫峰寺)、悯忠寺、天宁寺、胜果寺、宝集寺等。唐代寺庙由于年代久远，至今已没有留下任何木结构建筑遗迹，现在只能依靠保留在寺院中的石刻、经幢、石塔的唐代造型和石刻文字来了解唐代寺庙的位置和规模。

唐代幽州重要的历史遗存是北京房山云居寺的石经。佛教在中国的发展，曾屡遭打击，历史上的北魏、北周时期就发生过烧寺庙、毁佛像和焚烧佛经的事件。从隋代开始，僧侣们为使佛经能传于后世，就开始了开凿石窟，刻写石佛经的活动。隋代佛教僧侣静琬来到今房山白带山下，发现山腰上有许多大石洞，决定在此地把佛教经文刻于石窟壁上。到唐代初年静琬主持的刻经工程已颇具规模，刻经的同时静琬又在白带山下创建了云居寺。静琬还在山侧石壁上开凿石室，将刻好的佛经石板放进去，并将洞口封好。云居寺的刻经工作，自隋代名僧静琬开始，到唐末为止300余年间，共刊刻出石版佛经100余部，经板4000余块，都封存在白带山的洞窟中。

## 二、依水而建的都城——辽金北京的城市与建筑

唐朝后期藩镇割据，日益衰微，导致了中国进入五代十国的分裂割据局面。五代时期，后唐政权将北京地区在内的燕云十六州割给契丹。契丹是中国北方的一个游牧民族，祖先为东胡人，后为鲜卑族的一支，最初生活在潢河(今西拉木伦河)流域，后不断发展壮大。938年，契丹占领北京地区的幽州后，将幽州升为幽州府，定为契丹政权的陪都，称为南京。947年，契丹政权建国，国号辽，幽州成为辽南京。辽开泰元年(1012年)，将幽州府改名为析津府，辽南京更名为燕京。辽统治实行五京制，京下置道，道下置府、州、县。今北京地区分属南京、西京和中京三道管辖。北京城为南京道析津府所辖，南京析津府的城垣仍沿袭唐幽州城，未有大的改变。有关辽南京析津府的四至和规模仅能从史料和考古遗址中考证。

辽南京城的总体位置在今宣武区的广安门一带。在白云观墙后的工地上发现古城墙遗址，经考古学者确定为辽南京城北墙，东墙在今宣武区烂缦胡同和法源寺之间，西墙位于今会城门附近，今莲花河考证为辽南京的西护城河，南墙在今宣武区白纸坊路。辽南京城久为军镇，所以城墙高大坚固，据史料记载高达3丈，宽1丈5尺。城的总周长相当于现在的24里。辽南京城内的西南部建有富丽堂皇的宫城，据史料记载宫殿楼台巍峨，但至今未发现考古成果。

另据史料记载，辽代南京寺庙林立，庵院遍布。但目前发现的辽代寺庙遗物并不多见，只是在北京地区文献中保留许多有关辽代寺院的碑文、幢记，可知南京城内佛寺盛况。悯忠寺(今法源寺)始建于唐贞观十九年(645年)，据传为唐太宗征高丽回兵幽州，为悼念阵亡将士所建。辽代悯忠寺成为京城最大的寺庙，辽国皇帝常在此作佛事并接见外国使者。这里还是当年佛经研究、刻印和发行中心，山西应县木塔出土的许多辽代佛经大部分是悯忠寺印刷的[①]。现在悯忠寺位置为清代后期的法源寺，所留辽代遗物不多，但从寺庙布局来看应为辽代布局形式，仍可显出当年繁荣景象。辽南京城内至今遗留的唯一建筑实物就是位于北京宣武区西便门桥西的天宁寺塔了。辽代在城外郊区也分布许多著名的寺院，后来沿袭成为北京今天的著名风景区。较著名的有妙应寺(今白塔寺)、慧聚寺(今戒台寺)等。

1115年，女真族发展壮大并建立金国。天会三年(1125年)南下灭辽国。贞元元年(1153年)金朝将国都从上京会宁府迁都燕京，改名中都。从此金朝政权与南宋王朝对峙近百年，从这种意义上讲，北京的建都史为854年。

金朝行政区域统治沿袭宋朝和辽朝旧制，以路为一级行政单位统辖州府，州府下辖县镇。在金中都地区设置中都路，近郊部分设置大兴府。现今北京地区分别由中都路、西京路和北京路管辖。

在金朝政权把都城迁到中都之前，金海陵王天德三年(1151年)，金朝政权决定扩大燕京城的规模，将燕京城南、东、西三面城垣各向外扩展三里地，北城垣依旧。经学者根据考古推测，扩建后的金中都城，西南角在今丰台区凤凰嘴村西南，西北角在今军事博物馆南的皇亭子，东南角在永定门火车站西南四路通，东北角在宣武门内翠花街。城垣周长37余里(实测为18690米)。沿城垣有13个门：南城墙正中为丰宜门、左边为景风门、右边为端礼门；北城墙正中为通玄门、左侧门叫会城门，右侧有两个门，崇

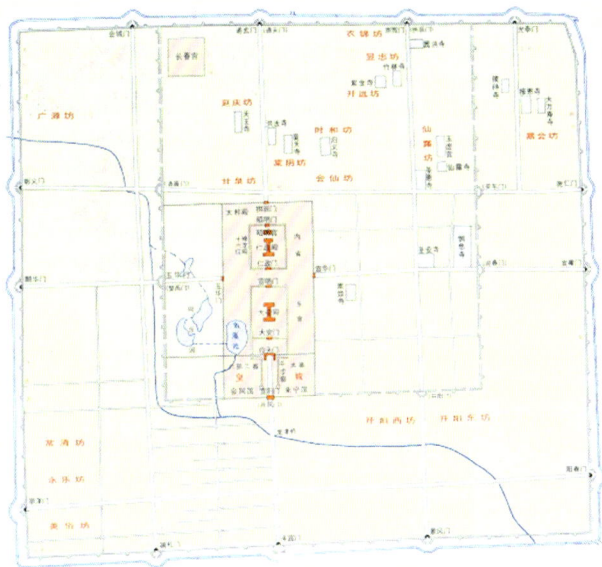

金中都平面图(摘自《北京历史地图集》)

智门和光泰门；东城墙正中的门叫宣曜门、左边为施仁门、右边为阳春门；西城墙正中叫颢华门、左边门叫丽泽门、右边叫彰义门。每座城门上都有城门楼。

金代的建筑技术，达到了较高的水平，中都城则集中反映出金代在建筑上的成就。金朝灭北宋攻陷汴京时，将大批建筑能工巧匠带到燕京定居，他们把中原的先进技术带到北方。在金中都的建筑物中，宫殿建筑水平最高。据史料记载金朝的宫殿群在辽南京宫殿群基础上吸收了北宋王朝宫殿的规制建成。金皇宫的最大建筑是大安殿，面阔11间，建在三层台基之上，史书记载高大庄严，惜至今未留有实物。上世纪90年代初，北京建设西厢路，西厢路由北向南贯穿金中都城的宫殿区遗址。经发掘基本确定了大安殿的位置。大安殿遗址位于白纸坊西街与滨河西路交叉处以北的一处夯土上，南北约长70米，东西约长60米，从遗址考查证明了史书的记载①。

金朝皇陵位于北京房山区云峰山下。据文献记载，金海陵王于贞元元年(1153年)定都中都，并于贞元三年(1155年)在云峰山下建陵，并将其祖陵迁来。后在其统治中国北方的120年中历经的9个皇帝全部葬于金陵中，故金陵葬有17个皇帝，陵区面积方圆160里，规模很大。明朝末年，由于东北与金同源的少数民族女真人强大，不断侵扰明朝边境，威胁明朝统治，为切断女真人的所谓"龙脉"，明天启二年(1622年)派兵对金陵进行了毁灭性破坏。清军入关后，对金陵进行部分修整。2002年6月，考古工作者对金陵

主陵进行了考古勘察和试掘，先后发掘并清理了主陵区的石桥、神道、台址、大殿遗址，并出土了大量建筑构件和随葬品。

金代现存最著名的一处遗迹就是著名的卢沟桥，初建于金大定二十九年(1189年)，完成于明昌三年(1192年)，桥长266.5米，桥栏杆上有501只石狮子(也有一说502只)，桥下11孔。这座成功运用拱券技术建造的石桥至今仍然基本完好。另外金代还保存了一些古塔等古建筑，如位于丰台区长辛店云岗村的镇岗塔、位于门头沟区雁翅镇淤白村北八里白瀑寺内的圆正法师塔以及昌平区的银山塔林等。

金中都南城垣水关遗址位于北京丰台区右安门外玉林小区，今凉水河以北70米处。水关又称"水门"、"水涵洞"，是河道沟渠穿过城墙的通水口。现存遗址是大块石条平铺河床，两边石条沿岸铺装的建筑遗迹，在石条下用沙石和柏木桩做基础，这种做法一直沿用到近代以至现代。水关遗址的发现确定了金中都城垣南城墙的确切位置和金代护城河的水源路线，证明了史料记载金中都水源南墙出城的准确性。

现存金中都城垣遗址是唯一确定金中都城墙位置的文物古迹。遗址位于丰台区和宣武区西部，现为6处城墙夯

金陵全景

土。这6处夯土位置是丰台区卢沟桥乡内保留有金中都西南城墙遗址3处；三路居凤凰嘴村为金中都城西南角一处夯土；万泉寺保留金中都南城墙两段；东学头高楼村有金中都西城墙一段。

## 三、严谨壮美的国际化大都市——元代大都的城市与建筑

元朝是蒙古族建立的一个皇朝。蒙古族原是漠北的一个游牧部族，1206年元太祖铁木真继位大汗，为成吉思汗。1234年元太宗窝阔台灭金，1271年元世祖忽必烈建立元朝，次年将金朝中都改名元朝的大都。1279年元朝灭南宋统一了中国。

元大都位于金中都城的东北郊，始建于元世祖至元四年(1267年)，至元十一年(1274年)宫殿区完工，至元十三年(1276年)元大都城基本完工，后屡有兴建。金中都以前，北京城的变化不大，基本上都延续了燕以来的蓟城的规划，但是元大都却是在金中都故城的东北方，以金代行宫大宁宫(今北海公园)为中心，营建一座新都城。原因有四：一是金中都城已于1215年被成吉思汗的大军摧毁，无法利用；二是大宁宫未遭破坏，忽必烈数次前来燕京时，都曾在琼华岛上的广寒殿暂住；三是金中都故城缺少充足的水源，当年金中都的莲花池和莲花河水系已不敷使用，而新城址可以利用玉泉山和高梁河的水系，并且著名水利家郭守敬已在1262年着手规划此事；四是另建新城，可以不受旧城束缚，能按理想放手规划。

元大都是继隋唐长安城之后又一座经严密规划建设的帝都，其规划之严谨、规模之富丽、建筑之壮美是中国最著名的三座帝都之一，在世界城市史上享有极高声誉。

在述及元大都地理位置时，元人陶宗仪在其《南村辍耕录》中写道："右拥太行，左注沧海，抚中原。正南面，枕居庸，奠朔方"是一个建都的好地方。在忽必烈即位之初，汉人理学家郝经就劝其将政治中心南移，说："燕都东控辽碣，西连三晋，背负关岭，前临河朔，南面以莅天下"(《郝文忠公集·班师议》及《便宜新政》)。南宋大儒朱熹从"风水学"角度也对北京做过论述，认为是个建都的好地方，他说："冀都正是个天地中好个风水，山脉(太行山)从云中(晋北)发来，前则前河环绕、泰山耸左为龙，华山耸右为虎，嵩(山)为前案，淮南诸山为第二重案，江南诸山及五岭为第三重案，故古今建都之地莫过于冀。所谓无风以散之，有水以界之也"(《朱子全书·地理》)。可见北京在古人的眼中既是一个"风水"宝地，而且在地理位置上也处于极其重要的地位。

刘秉忠在相当于今北京鼓楼的地方，建立了一座中心台，以此为四面城墙的定位提供标准。他规划的大都城为南北向略长于东西向的长方形，周回60里，面积约为金中都城的2.7倍。共开有11个城门，城门相对之间都有通衢大道，即《考工记》说的"九经九纬"，纵横街道所分隔成的方格地块，即为"坊"，坊内是居住区或衙署区。大都城共有50个坊，每个坊都有吉祥的名称，如今天北京阜成门内大街东部地区，在元朝就属"金城坊"。忽必烈曾下诏规定，原住在金中都旧城内的有钱人家和官员，可以优先在大都城内各坊领取地块，每家8亩，建造住宅。不准超标准占地，也不准占着地不建房。不建房的人家，要把地块让出来交别人建房。这个诏书应是北京城里关于房地产规定的最早的文件。明代《永乐大典》中收录的元末熊梦祥所著《析津志》中说，"大都街制，自南以至于北谓之经，自东至西谓之纬。大街二十四步阔(一步约合1.54米)，小街十二步阔，三百八十四火巷，二十九锡通"。锡通即是今天说的"胡同"，出自蒙古语。

规划布局上，元大都最接近《考工记》所提出的理想，以中国封建社会最高统治者所追求的理想秩序为蓝本，即方正的城市外廓，以贯穿全城南北的中轴线为对称轴的东西对称格局，皇城、宫城位于全城中轴线上的显赫地位，严格的纵横正交的街道网格，左面建设"祖"、右建"社"，宫前面建广场，后面有市。其规划技术上先确定南北中轴线，依北海之北的什刹海东岸最突出的一点(现地安门之北后门桥处)而定，在此点稍北即今北京鼓楼的地方，确定为全城几何中心，建"中心之台"；再依其与金中都北墙的关系确定南墙，以中心台与南墙的距离为半径确定北墙；西墙则依什刹海迤西，当时积水潭之西端而定，把积水潭水面全部包入城内，再以中心台与西墙之距为半径确定东墙，只是因施工时发现东墙正在洼处，故稍向内移，影响到今北京中轴线与东二环路的距离稍短于与西二环路的距离。在后门桥河泥里曾发现石鼠(按地支，子方为北方即所谓子鼠)，在前门附近河里曾发现石马(按地支，午方为南方即所谓午马)，应该说这就是北京中轴线(子午线)的标记物，而且有关学者研究这条中轴线直通千里之外蒙古上都城，与上都城中轴线连成一体。大都皇城布置在城市中轴线上南部的中央，皇城的中轴线(与明清北京中轴线同)，上建宫城，又称大内，正门称崇天门。皇城之北鼓楼一带

是最主要的市场。在皇城外左右，都城东西城门齐化门和平则门内，分建太庙和社稷坛。皇城南部偏东为宫城，大明殿为前朝，延春阁为后宫。宫城西为太液池，就是今天的北海和中海（当时南海还没有开辟），其西岸建有隆福宫和兴圣宫。积水潭北岸到钟鼓楼一带为繁华的商业区。这些都明显与《考工记》的规定相符。

建成的大都城平面长方形，东西两面城墙略长于南北两面的城墙，东西6700米，南北7600米，四周城墙长约57里，史载方圆60里，经勘测为28600米，面积约51平方公里，远大于金中都，约与唐洛阳城相仿。元大都城的城墙，全部用夯土筑成。大都城辟有11个门：南城墙为顺承门、丽正门、文明门，北城墙为健德门和安贞门，东城墙为光熙门、崇仁门和齐化门，西城墙为肃清门、和义门和平则门。门的名字都语出《易经》而得。这些门现已无存，仅在上世纪70年代拆除北京城明清城墙西直门城楼时，在西直门瓮城城墙内包着一个完整的元大都和义门城门，为现代留下唯一的元大都城门遗迹。据对遗址一段完整的城墙进行测绘得知，元大都城墙最下端宽约24米，城高约16米，顶部宽约8米。据对元大都和义门的勘察得知，元大都每个城门上都有箭楼，箭楼面阔三间，进深三间，地面用砖铺砌，旁边有砖砌水池，是为防御火攻的设施。在城墙的上端，沿中心线部分，还铺设了半圆形瓦管作为排泄雨水的设施。元大都城墙因为是用夯土筑成，所以在建成之初被暴雨冲刷多有坍塌。为保护城墙，元朝政府专门在文明门外设立芦苇场，征收芦苇，编成芦苇衣挂在城墙上防

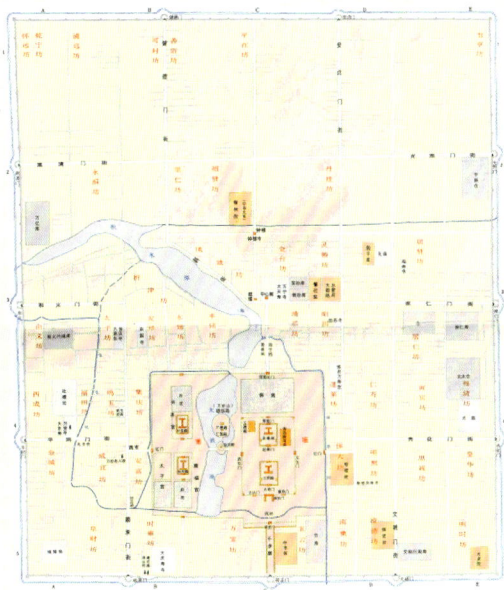

元大都平面图（摘自《北京历史地图集》）

护城墙。到以后城墙不断被雨水冲后加固夯筑，已十分坚固，不再用苇衣保护。现在元大都城墙遗址仍有很长一段保留，它的位置在今天的西直门以北学院路和西土城路交汇处向北延伸，在学院路向东延伸至朝阳区太阳宫，虽经700余年的雨水冲刷和人为破坏，至今保留数米高的一座笔直的土山，这座土山清晰地反映出当年城墙的轮廓，从坍塌土层的断面显示出多层夯筑的痕迹。在元大都城墙遗址的塔院段，有一处保留完整的水涵洞，砖石券洞和铁栅栏仍保留完好。

另外，水系的成功规划是元大都的又一大特色。元大都的水系是由杰出的科学家郭守敬规划的。他一方面疏通大都东面的运河——通惠河，同时又规划了一条新渠，从今天昌平的白浮泉引水，延北部山脚修渠向大都供水，汇合西山泉水，在北城汇成湖泊，然后通入通惠河。这条新渠汇合了大量的水源，既解决了大都的用水，又开通了运河，使南方物资可以通过运河直达大都，解决了大都的粮食等物资运输问题。

马可·波罗在其游记中称赞大都道"街道甚直，此端可见彼端，盖其布置，使此门可由街道远望彼门也。城市中有壮丽宫殿，复有美丽邸舍甚多"，"各大街两旁，皆有种种商店屋舍，全城中划为方形，划线整齐，建筑屋舍……方地周围皆是美丽道路，行人由斯往来"。"其美善之极，未可言宣"（《马可·波罗行记》）。

为了维护元朝政权的统治，元朝政权对各种宗教派别采取宽容和利用的态度，除了对原有的佛教广泛传播，道教、正一教、藏传佛教同时发展，建造许多大型庙宇。如大都的护国寺、妙应寺、东岳庙等，虽经后代改建，从现存的遗迹中仍能看出元代庙宇的遗迹。例如位于朝阳门外的东岳庙，从主殿的周围廊、工字殿和大殿二殿台明纵向连通的形制，仍保持元代及宋、金庙宇的建筑特色。尤其是东岳庙大殿东配殿的木构架，从摆放稀疏和造型硕大的斗拱中读出了元代建筑的特点，许多专家以为这栋建筑木结构留有较多的元代遗物。京郊门头沟清水镇齐家庄村的灵严寺仍存有元代遗构。藏传佛教至今保留较多的是覆钵式塔，现存最著名的便是西城区阜成门内大街的妙应寺白塔。过街塔是元代藏传佛教的另一种建筑类型，以北京昌平居庸关的云台最为著名，现塔已无存。

元代大都经济繁荣，人口众多，一般民宅为四合院，元代四合院的造型已成熟。元代四合院目前在北京已无实物，唯一能供参考的就是元大都旧址发掘出来的位于北京

西城区后英房胡同元代住宅遗址，根据遗址可知元代民居已形成正房、厢房四合而围的建筑形式，院落分布在胡同两边，形成城市建筑最基本的单元。

## 四、中国封建帝都的典范——明清北京的城市与建筑

元末农民起义推翻了元朝的统治，明太祖朱元璋于1368年建立了明朝。明朝原定都南京，1403年明成祖朱棣获得帝位后，为防御蒙古残余势力的南侵，把政治中心与军事中心合并，决定把明朝的首都迁到北京。北京的建都工作从明永乐五年（1407年）开始，到永乐十九年（1421年）正式迁都北京，北京改称京师，习惯仍称北京。明朝的北京城是继承历代都城建设经验基础上创造出来的，是中国

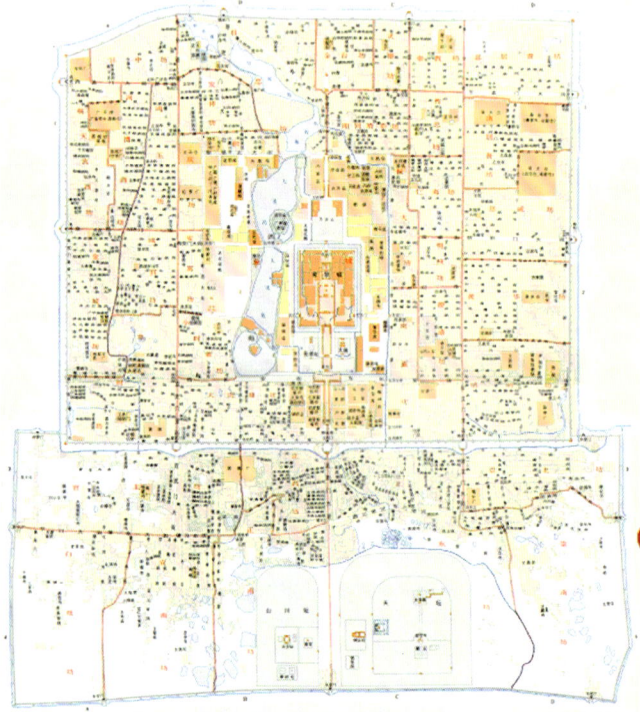

明朝北京城平面图（摘自《北京历史地图集》）

封建帝国都城建设的典范，集中体现了历代都城建设的精华，在设计思想、规划布局、建筑施工上都达到了封建社会的最高水平，也是全国古都中保存最好、规模最宏大的一座。迄今为止，北京完整地保留了明代北京城的基本格局和很多明代建筑。

明代在设计国都北京时，仍运用元大都原有格局，如宫殿设置、街道规划仍大部分用元大都基础，同时仿照南

京的建制和历代王朝都城的规划思路，兴建一座新的宏伟的都城，此后一直沿用至清朝灭亡。

明代北京城墙在元代土筑城墙的基础上，出于防御的需要放弃元大都城北部的城市建设预留地将城墙向南退5里，将南部城墙从今长安街一线南扩800米左右至今宣武门崇文门一线，并在外层用大城砖将土城包砌，形成长方形城垣。全城东西距离为6650米，南北距离为5350米。南城墙设三门，中间为正阳门、东为崇文门、西为宣武门；东城墙设两门，东直门和朝阳门；西城墙设两门，西直门和阜成门；北城墙设两门，德胜门和安定门；城四角建筑高大的角楼。城市中轴线仍然沿用元代的中轴线，城市布局沿着中轴线对称展开。皇城位于内城中心向南偏一点，东西长2500米，南北长2750米，呈不规则方形，为避让元朝大庆寿寺使西南角凹进。皇城四向开门，南向为承天门，承天门前面还有皇城的前门，明代称大明门，清朝改为大清门。大明门和承天门之间建筑千步廊，千步廊东西建文武衙门等官署。皇城北向为北安门，东向为东安门，西向

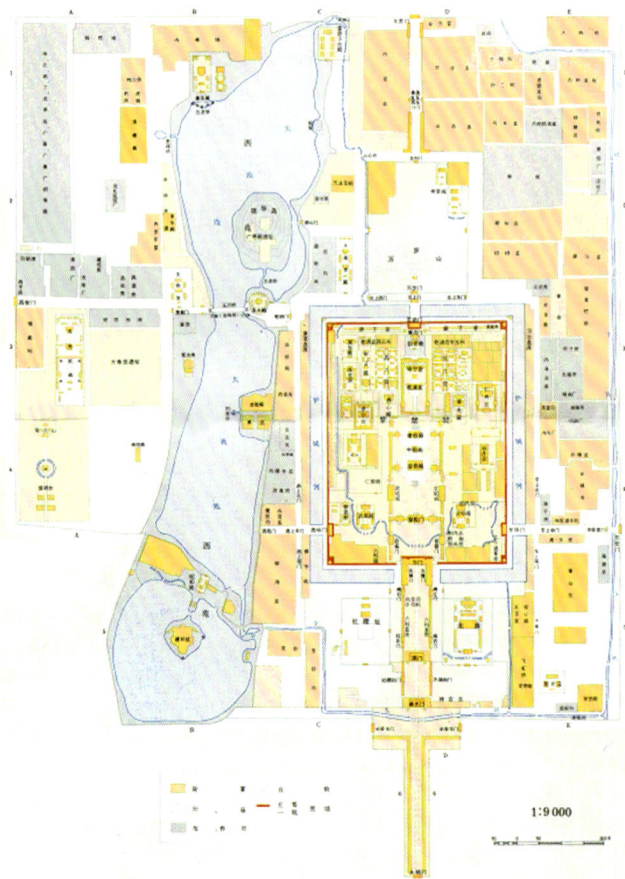

明朝北京城皇城平面图（摘自《北京历史地图集》）

为西安门。皇城内主要建筑为宫苑、寺庙、衙署和仓库。皇城中央安排最重要的建筑宫城，即紫禁城，也就是今天的故宫，紫禁城南北长960米，东西宽760米，四面有高大的城门，南为午门，北为玄武门，东为东华门，西为西华门，城四角建有形制华丽的角楼。宫内主要建筑前三殿和后两宫都建筑在城市中轴线上，宫殿后面是御苑，御苑内堆砌高大的万岁山作为背景(亦名镇山、景山)是拆除元代宫殿时的废土和开挖紫禁城护城河的土堆积而成，正好压在元代宫殿延春阁之上，以示镇压前朝之义。明代北京城更加强调以皇城为中心和突出轴线，首先城市重要的高大建筑多数集中在中轴线上，前面从正阳门开始向北经过宫殿区，而且为了更突显皇宫的辉煌和至高无上，宫殿区全部采用黄琉璃瓦，而且规定民宅只能用灰瓦，高度也有限制，因此在全城一片灰瓦矮房中宫殿区显得更加耀眼夺目。宫殿区向北为全城至高点景山，再向北以高大的钟鼓楼为轴线收尾，整条中轴线得到了极大的加强。明嘉靖三十二年(1553年)为加强北京城防加筑一个外城，但外城在

南部建成后向北延伸的财力不够，因此在北京内城南城墙封护，形成南城外城垣。外城的城墙和城楼的规模都比内城要小一些。北京城外城东西长7950米，南北长3100米，南城墙置三门，中间为永定门，西侧为右安，东侧为左安；东、西、北向各两座门，西侧为广安门和西便门，东侧为广渠门和东便门。这些城门都有瓮城、城楼和箭楼，城墙上有士兵驻守的值房。由于外城的加筑，原来位于城外的天坛和先农坛也被包进城内，永定门直通正阳门，使得北京的中轴线再次向南加长达永定门，形成以城门为先导，以两座宏大的坛庙为前翼，以恢宏的宫殿为高潮，以高大的钟鼓楼为收尾的全长7.8公里的城市中轴线，成为古今中外最长的一条城市中轴线。另外城市中所有重要的宫殿建筑全布置在中轴线上或者围绕中轴线两边。这条轴线以外城南端的永定门为起点，向北延伸，在轴线上的建筑有永定门、正阳门、大明门、承天门、端门、午门、皇极门、皇极殿、中极殿、建极殿、乾清宫、坤宁宫、玄武门、万岁山、北安门，以鼓楼、钟楼为轴线终点；在中轴线两侧布置天坛、先农坛、太庙、社稷坛、中南海、北海。这种布局形式使皇家建筑占据全城的中央部分，源于封建古代的礼制，也便于统治的需要，但妨碍了全城东西方向的交往。

北京内城的街巷，大体沿用元大都的规划，分布在中轴线皇宫衙署的东西两侧。与正阳门并列的崇文门、宣武门，从两门向北各有一条大道达北京内城北部，与东直门、西直门、阜成门、朝阳门横向的四条大街相交，北京的街道和胡同全相交于这六条道路。这些道路平面构图上相互垂直相交形成方格形，是中国古代城市街道布置的代表作品。在大小干道两侧是商业和手工业，胡同小巷为四合院居住区。

另外，在北京昌平区天寿山麓建造了明朝13代皇帝的陵墓，一般称为十三陵。十三陵陵区北、东、西三面由山岭环抱，南面临平原，十三陵墓组群分布在山谷中。明朝迁都北京的第一个皇帝明成祖的陵墓——长陵居中心位置，其他12个陵各依地势分布，彼此相距四五百米至1000米不等。在陵区山谷前的缓坡上，距长陵约6公里为陵区入口，入口处有一座巨大的石牌坊作为标志，石牌坊北为陵区大门，大门向北排列碑亭、华表、龙凤门和18对巨大整石雕刻的文臣、武将、大象、马、骆驼等雕像。长陵建成于明永乐二十二年(1424年)，它是十三陵中最大的一座。长陵是由宝顶、方城明楼和位于明楼前的祭殿祾恩殿组

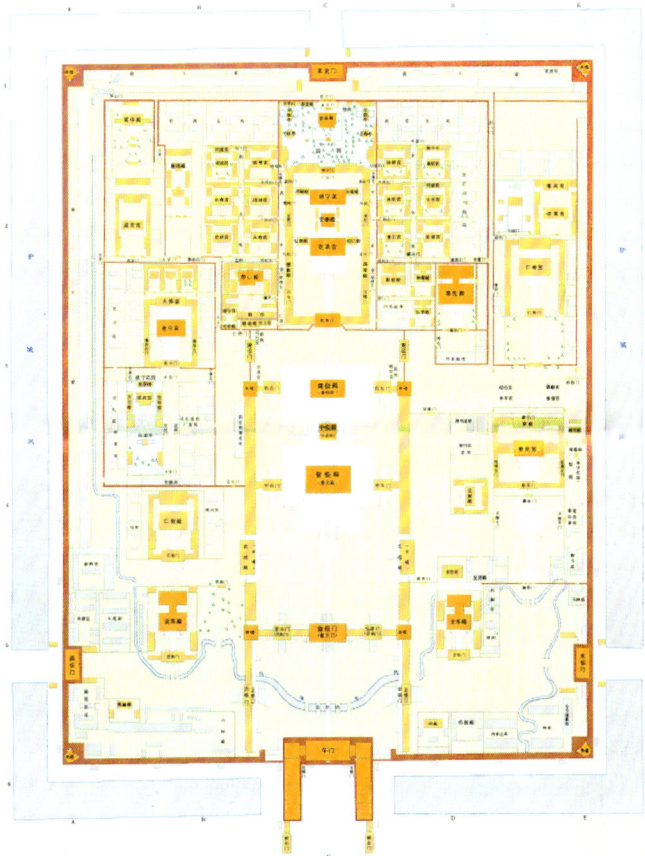

明朝紫禁城平面图(摘自《北京历史地图集》)

成。宝顶为墓上封土，封土周围用圆形的城墙围护，封土下为深埋于地下的地宫。宝顶前面正中部分做成方台，上立碑亭，方台称方城，碑亭称明楼。长陵祾恩殿是一座和皇宫中太和殿相类似的大殿，面阔九间，重檐庑殿顶，下面由三层石台基承托，面积也和太和殿相当。祾恩殿内部使用32根整根优质楠木柱，最高的约12米，中央明间的4根大柱，直径达1.17米，是现今古代遗物中的精品。

明朝陵墓地下墓室都是用巨石发券构成若干墓室相连的"地下宫殿"。1956年，考古工作者发掘了定陵的陵墓，供游人参观。现存十三陵地上建筑多为明朝所建，有少部分为清代复原。

北京现存明代建筑较真实和完整的就是蜚声中外的万里长城。明朝是推翻北方少数民族元代政权而建立的朝代，明代的首要任务就是防止元代政权卷土重来，修筑长城成为整个明朝政权的重中之重。自从明洪武元年(1368年)徐达攻克元大都后，就开始修筑关隘、城墙和烽燧，此后，几乎每一个明朝皇帝均把长城的修筑当成大事。长城主体约在1600年左右基本完成。长城的主体是城墙，城墙的位置多选择在山脉的分水线上建造，其构造按地区特点有条石墙(条石内夯筑夯土或三合土)、块石墙、夯土墙、砖墙等建筑构造。凡长城经过的险要地带都设有关城，关城是军事通道，防御设置极为严密，重要的关城形成军事重镇，如北京的居庸关是京师北部的通道。

明代保留至今的建筑还有北京先农坛太岁殿、智化寺和天坛神乐署等。

1644年，明朝被李自成领导的农民起义军所灭亡。同年，身居东北的满族入关，占领了北京，并建立清朝。满族是居住在中国东北长白山一带女真族的一个部族，1616年自称后金，1636年改称清。

顺治元年(1644年)，清朝建都北京。北京又称京师，地方政区建制基本沿袭明朝制度。省为一级政区，下辖府、州、县。北京设直隶省，下辖顺天府、承德府、宣化府等11个府，遵化等6个直隶州。光绪三十四年(1908年)，北京城为顺天府管辖，下设5州19县，大多地处今北京市境内。除局部建制和市政管理有所变化外，北京城内总体格局和街道系统仍同明代。局部建制有如下改变。明代皇城前为中央官署集中区，清朝废五军都督府，在皇城千步廊右侧的原明代五军都督府旧址建旗人住房，改变了中央官府沿千步廊左右布置的局面；废弃了明皇城内的各类仓库和厂房，建成官府和民居，打破了明代皇城的封闭局面。

在市政管理方面，清初在内城废除了明代的坊铺制管理，由八旗军带家属按方位管辖和居住，而汉人都居住在外城，主要是中轴线南端，前门以南两侧。

清代皇城沿用明代皇城，仍设四门，但在明代皇城基础上作了一些改变。北安门改为地安门，承天门改为天安门，大明门改为大清门。万岁山改为景山，俗称煤山。乾隆十五年(1750年)在景山建造5个亭子，其中位于全城中轴线上的万春亭体量最大，西侧为观妙亭、周赏亭，东侧为辑芳亭和富览亭。清代将原明代北海和中南海的宫殿多废除和改建，变化较大。琼华岛上的白塔为顺治八年(1651年)修建，并在塔前广寒宫旧址上增筑永安寺。

清代紫禁城仍沿用明代所遗留的建筑格局，但明紫禁城建筑大部分在明末战乱中焚毁，顺治初年又在旧基础上重建，至康熙二十五年(1686年)基本完成。后因是皇家所在，资金充足，屡屡修建增建，至嘉庆年间规模成型，因此紫禁城内明代建筑几乎都经过改建，明代建筑遗存均消失。紫禁城北门明称玄武门，清改称神武门，余下三门仍按明代为午门、东华门、西华门。清紫禁城格局仍沿明朝，主体建筑仍为"前朝"办公，"后廷"为生活区。

清朝紫禁城三大殿改称太和殿、中和殿、保和殿。太和殿为朝会之所，即举行国家重大礼仪盛典的场所。中和殿为接见内阁和各部大臣的场所。保和殿为宴请外宾、举行殿试的场所。前朝两侧文华殿、武英殿东西并列，仍用明代称呼。后廷三大宫沿用明代称谓和功能，反映出清朝政权遵从汉族政权传统的做法，一反中国历代政权更迭废弃前朝一切的习惯。乾清宫仍为召见大臣场所，交泰殿存放御玺，坤宁宫为帝后寝宫。后廷两侧东西六宫仍为帝后居住。

清代北京建筑组成元素上也发生了一些变化，各种建筑类型更加丰富。首先，清代统治者不习惯在紫禁城高大森严的宫殿建筑里办公和生活，历代皇帝都热衷于在京郊山水间建行宫别业，并长驻行宫游乐和处理政务，因此行宫规模不断扩大，形成了清代北京西山的皇家园林——三山五园。三山五园均分布于京郊西部，分别是香山静宜园、玉泉山静明园、万寿山清漪园、圆明园、畅春园。其中规模最大的是圆明园。圆明园始建于清康熙四十八年(1709年)，雍正皇帝扩建为皇家别墅，乾隆皇帝再度扩建，于乾隆九年(1744年)竣工，后在圆明园旁建长春园和绮春园，到乾隆三十七年(1772年)建成，成为今日遗址展现规模，通称圆明园。

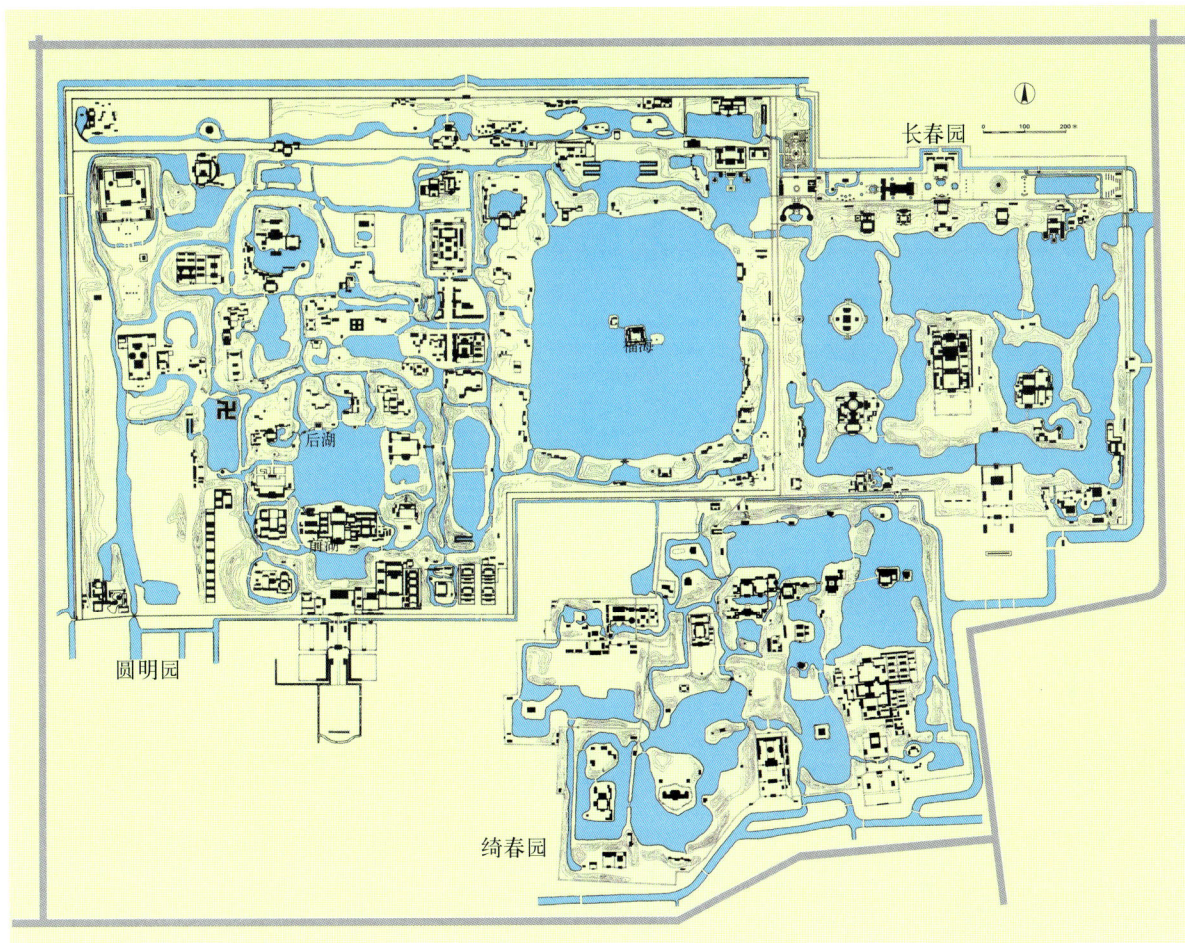

圆明园遗址平面图

其次，清政府为团结蒙藏民族，在北京兴建了众多的喇嘛教寺院，至今保存完好的有嵩祝寺、福佑寺、妙应寺、护国寺、西黄寺、雍和宫等。这些喇嘛寺院的建筑形式，没有按照蒙藏地区喇嘛寺庙的建筑形式和格局建造，而是完全按照佛教寺院的建筑布局和建筑形式，仅在大殿内外部装饰喇嘛教风格的装饰，说明清政府对喇嘛教的态度仍是以中原文化为中心。

再次，清政府规定全部宗室都集中在京城建造府邸，不分封各地，因此京城的王府建筑也成为一个独特的建筑元素。清朝的王府建筑规模和形式仅次于皇宫建筑，以其宏大的规模、严整的规划布局、多重院落和华丽的彩绘、琉璃装饰在北京传统建筑中占有特殊的地位。

最后就是北京城大大小小的四合院住宅了。据清乾隆十五年（1750年）绘制的《乾隆京城全图》记载，当时北京城内格局完整的四合院约有4万多个。四合院是北京住宅的主要形式。到清代已形成固定建筑格局。这种格局就是将成排的房子四面围合，称为四合院。四合院的四个角通常用走廊或院墙将建筑连接起来，成为一个既安全，又可防风沙的封闭的院落。住户在封闭的庭院中种植花木，喝茶聊天，形成一个安静舒适的生活环境。

建国后经过近50年的发展，北京已成为一个现代化都市，古老的建筑和现代风格建筑交相辉映，保护和发展是人类永远的课题。

侯兆年
2007年2月

注释：
①北京市文物研究所、平谷县文物管理所上宅考古队，《北京平谷北埝头新石器时代遗址调查与发掘》，《文物》1989年第8期。
②北京市文物研究所，《琉璃河西周燕国墓地》，文物出版社，1995年版。
③北京市文物管理处，《北京外城

东周晚期陶井群》，《文物》1972年第1期。
④王世仁：《北京天宁寺塔三题》，《北京文博》1996年第2期。
⑤赵福生：《金中都考古工作取得重要进展》，《北京考古信息》1991年第1期。

# An Outline of Ancient City of Beijing and Its Architecture

Beijing, as the capital of the People's Republic of China, is the center of politics, economy and culture of this country. With a long history and numerous cultural relics, Beijing is world's famous for its status as an ancient capital and for its history and culture. As early as over 3,000 years ago, a city emerged in Beijing, as evidenced by the site of the Shang and Zhou Dynasties found in Liulihe, Fangshan District, the splendid Yuan Dadu (the Capital of the Yuan Dynasty), the capital of the Ming and Qing Dynasties, magnificent palaces, mysterious temples, beautiful gardens, the grand Great Wall, solid city defense system, solemn imperial tombs, quadrangles etc., all of which constitute the most complete and the most essential cultural heritage typical of Chinese ancient architecture. In Beijing, there are 6 world cultural heritages, 98 national key relics under special preservation, over 200 municipal relics under preservation and over 500 district or county level relics under preservation.

## 1. Formation and Development of the City: Beijing's City and Architecture from Prehistory to the Tang Dynasty

One of representative sites of the Neolithic Period discovered in Beijing area is the tomb group of Donghulin People. These tombs are located at loess mesa on the western bank of Qingshuihegu Valley on the western side of Donghulin Village, Mentougou District, in the western suburb of Beijing. According to archaeologists' judgment, the Donghulin People were dated back about 10,000 years, the early Neolithic Era. Human sites of this period also include Shangzhai and Beiniantou sites at Pinggu County, Zhenjiangying site in Fangshan District, Xueshan site in Changping District etc. Beiniantou site is situated about 7.5 kilometers northwest of Pinggu County. 10 resident sites of the Neolithic Era have been discovered there. These houses are round or elliptic half-vaults with a diameter of generally about 4 meters. It is estimated that this site was built 6,000 to 7,000 years ago. As the earliest human living

site discovered in Beijing area to date, the site provides valuable materials for the research on Beijing's early architecture.

In 1984, archaeologists discovered a large-scale site of the Shang and Zhou Dynasties at Dongjialin Village, Liulihe Township, Fangshan District, Beijing, known as Liulihe site in academe. Scholars have agreed that this site was the initial feudatory of the State of Yan in the early years of the Zhou Dynasty. The site is 3.5 kilometers long from the west to the east, and 1.5 kilometers wide from the north to the south. It includes living district, tomb district and ancient city.

The State of Yan became a major Zhuhou State in the north of China during the Spring and Autumn Period, with Ji City being the capital. The city was near the White Cloud Taoist Temple and Guang'anmen in the southwest of old Beijing city. Today's northern mountainous area in Yanqing County and Huairou County, Beijing, was used as activity area of Shanrong Tribe. The Shanrong tomb groups discovered at places such as Yuhuangmiao in Yanqing County proved the activity scope of Shanrong Tribe.

The State of Yan was one of Seven Powers in the Warring States Period. The capital of the Yan State was still Ji City, the former capital in the Spring and Autumn Period, which is now in the southwest of Beijing. Besides Ji City, the State of Yan set 2 other capitals: Zhongdu and Xiadu. Zhongdu was situated in the west of Doudian, Fangshan County, Beijing. An earthen city site was discovered near Doudian in 1959, which is square and has inner and outer city walls. An earth cofferdam serves as the outer city wall, and the inner city is 1,230 meters long from the west to the east and 1,040 meters wide from the north to the south. According to the research, this city was Liangxiang City in the Han Dynasty or Zhongdu of the Yan State.

In the Western Han Dynasty, Emperor Wudi set Youzhou Cishi (Provincial Governor) Administration in

Beijing area. Ji City was renamed and called Youzhou the first time when the Eastern Han Dynasty was founded in 25. Beijing area was in the scope of Ji City, Yan area, in the Western Han Dynasty. According to the history records, Ji City had architecture of quite a scale. It had a southern gate leading to the capital avenue, as well as other gates with towers on. Kings of the Yan State had their palace city, within which there were Wanzai Palace and Mingguang Palace with brooks and pools beside the palaces, indicating a fairly large architectural scale.

A mausoleum of Prince Yan (or Prince Guangyang) of the Western Han Dynasty was discovered at Dabaotai Village, Fengtai District, Beijing, in 1974. It is in the shape of "凸", 23 meters long from the north to the south, 18 meters wide from the west to the east, and 3.7 meters deep, consisting of tomb dome, tomb way, corridor, inner and outer ambulatories, Huangchang Ticou (walls of yellow-core cypresses heading inward around the coffin), front chamber (for mourners to have a rest), back chamber (coffin room) and coffins etc. There are lots of funerary objects in the mausoleum, representing Beijing area's economical and technological development levels in the Western Han Dynasty.

Architecture in Beijing area had developed to quite a scale in the Western Han Dynasty. Over years many ancient city sites of the Han Dynasty have been discovered in the suburbs of Beijing, e.g. Guangyang County site at Guangyangcheng Village, Liangxiang Township, Fangshan District, Doudian earthen city site, Changgou earthen city site, Caizhuang ancient city site at Zhoukoudian, Lu Village ancient city site, Jundu city site in Changping District, and ancient city site at Beichengzi Village in Pinggu County. These ancient city sites are densely located, indicating prosperous economy at that time.

In the Eastern Han Dynasty, architectural relics in Beijing area were mainly tombs and stone pillars in tomb ways. A well-known is the stone pillar engraved with

"Deity Way for Youzhou Shuzuo (a job role) Qin Jun (the name) of the Han Dynasty", which was built in the 17th Yongyuan Year of the reign of Emperor Hedi of the Eastern Han Dynasty (105 A.D.), and discovered at Babaoshan in the western suburb of Beijing. Tombs of the Eastern Han Dynasty also had large sizes. For example, among 9 Eastern Han Dynasty tombs discovered in the north of Huairou County, Beijing, No. 31 multi-chamber tomb of the latter half of the Eastern Han Dynasty has the largest size. The architectural scale and development level can be estimated through these tombs. No actual building of the Han Dynasty exists today. However, so many ceramic house models as funerary objects must be models of houses in which tomb owners had ever lived. The ceramic house excavated from No. 31 tomb is very impressing.

Ji City (today's Beijing) in Youzhou was the capital of Youzhou in the Tang Dynasty, also called Youzhou City. Archaeologists, according to epitaphs of the Tang Dynasty discovered all the years and combining preliminary research of Youzhou City, have in recent years preliminarily determined the location and scope of the Tang Dynasty's Youzhou City. The eastern city wall was at a line from the north to the south linking today's Lanman Hutong and Fayuan Temple, Xuanwu District, Beijing. The western city wall was at a line from the north to the south linking Xiaomachang in the east of today's Huichengmen Village, Xuanwu District, Beijing, the eastern side of Beijing Steel Plant and the eastern side of Ganshiqiao. The southern city wall was at a line across the east and the west of today's Baizhifang Street, Xuanwu District, Beijing. The northern city wall was near today's White Cloud Taoist Temple, Xicheng District, Beijing. The Buddhism prospered during the Tang Dynasty, and the number of temples in Youzhou area increased continuously. Famous temples were Huiju Temple (later known as Jietan Wanshou Temple) in Ma'anshan, Mount Baidai Yunju Temple (today's Fangshan Yunju Temple), Yuni Temple (later known as Jiufeng Temple), Minzhong

Temple, Tianning Temple, Shengguo Temple, Baoji Temple etc. An important historical relic in Youzhou of the Tang Dynasty was stone slabs inscribed with Buddhist scriptures at Yunju Temple in Fangshan District of Beijing. Stone slabs carved at the temple, from the time of famous Master Jingwan of the Sui Dynasty to the end of the Tang Dynasty (totally 300 years), produced more than 100 sutra versions, and about 4,000 stone slabs, which were conserved in caverns in Mount Baidai.

## 2. Waterside Capital: Beijing's City and Architecture of the Liao and Jin Dynasties

Liao Nanjing (the Southern Capital of the Liao Dynasty) was near today's Guang'anmen, Xuanwu District. An ancient city wall site has been confirmed by archaeologists to be the northern wall of Nanjing City, which was discovered at a construction site behind walls of White Cloud Taoist Temple. The eastern wall was between today's Lanman Hutong and Fayuan Temple, Xuanwu District. The western wall was near today's Huichengmen, and today's Lianhua (lotus) River was the western moat of Nanjing of Liao Dynasty according to research results. The southern wall was at today's Baizhifang Street, Xuanwu District.

According to historical materials, there stood temples all around Liao Nanjing. The Minzhong Temple (today's Fayuan Temple) started to be built in the 19th Zhenguan Year during the reign of Emperor Tang Taizong(645 A. D.). It is recorded that Emperor Tang Taizong went on an expedition in Korea and then retreated to Youzhou. To memorize officers and soldiers that had died in the battlefield, he ordered to build such a temple. The Minzhong Temple became the capital's largest temple in the Liao Dynasty. Now, the temple is located at where the former Fayuan Temple was in late Qing Dynasty, with few relics of the Liao Dynasty left. However, the layout of the temple shows a style of the Liao Dynasty and we can still figure out prosperous scenes at that time. The only actual building in Nanjing City that exists today is the Pagoda at the Temple of Celestial Tranquility in the west of Xibianmen Bridge, Xuanwu District. There were also many famous temples of Liao Dynasty in the suburbs, which have shaped today's landscaped areas in Beijing.

Among them were Miaoying Temple (today's White Stupa Temple), Wisdom Accumulation Temple (today's Ordination Terrace Temple) etc.

In the 3rd Tiande Year of the reign of King Hailing of the Jin Dynasty (1151 A.D.), Jin administration decided to extend the scale of Yanjing City. Yanjing City's southern, eastern, and western walls were extended by 1.5 Kilometers respectively, and the northern wall remained unchanged. According to archaeological deduction, the extended Jin Zhongdu (the Central Capital of the Jin Dynasty)had the southwestern corner in today's southwest of Fenghuangzui (phoenix beak) Village, Fengtai District, the northwestern corner at Huangtingzi (royal pavilion) in the south of today's China Military Museum, the southeastern corner at Silutong in the southwest of Yongdingmen Railway Station, and the northeastern corner at Xuanwumennei Cuihua Street. The circumference of the city walls was about 18.5 kilometers (actually measured as 18,690 meters). Among the buildings in Jin Zhongdu, palaces featured the highest level. According to historical materials, palace complex of the Jin Dynasty was built based on that in Liao Nanjing and by referring to the palace specifications of the Northern Song Dynasty. The largest building of Jin Zhongdu was Da'an Hall, being 11 bays wide, built on a 3 tiers terrace. The history records said it was high and grand, but its actual relic doesn't exist today. The location of Da'an Hall has been basically found through the excavation. The site of Da'an Hall, now a rammed earth residue, located in the north of the intersection of Baizhifang Xijie and Binhe Xilu, is about 70 meters long from the north to the south and 60 meters wide from the west to the east.

The imperial mausoleums of the Jin Dynasty are at the foot of Mount Yunfeng (cloudy peak), Fangshan District, Beijing. According to documents, King Hailing settled down Zhongdu in the first Zhenyuan Year (1153 A.D.). He began building his mausoleum at the foot of Mount Yunfeng in 1155, and migrated his ancestral mausoleums there. 17 emperors were buried there, and the grand mausoleum district occupied an area of 6,400 square kilometers. The ruler ruined the mausoleums of the Jin Dynasty in the 2nd year of the reign of Emperor Tianqi of the Ming Dynasty (1622 A.D.).

The most famous relic of the Jin Dynasty is Lugou

Bridge, which was initially built in the 29th Dading Year of the Jin Dynasty (1189 A.D.) and completed in the third Mingchang Year (1192 A.D.). The bridge is 266.5 meters long with 501 stone lions (or 502 according to other sources) on the railing and 11 holes beneath. It, built successfully with arch technology, is still in a basically good condition. Besides, there still exist some ancient architecture of Jin Dynasty, e.g. Zhengang Pagoda at Yungang Village, Changxindian, Fengtai District, Stupa for Senior Monk Yuanzheng in Baipu Temple 4 kilometers north of Yubai Village, Yanchi Town, Mentougou District, and Silver Mountain Stupa Forest in Changping District etc.

The water gate ruins on the southern city wall in Jin Zhongdu are in Yulin sub-district, You'anmenwai, Fengtai District, Beijing, 70 meters north of today's Liangshui River. The discovery of the water gate ruins helps determine the accurate location of the southern city wall of Jin Zhongdu and water source route of the moat in the Jin Dynasty, proving the historical records are right in that the water source of Jin Zhongdu went out of the city through the southern wall. Existing city wall site of Jin Zhongdu is the only historic site that helps determine the location of city walls of Jin Zhongdu. The site is located in the west of Fengtai District and Xuanwu District, and has 6 rammed earth residues that were ever the city wall.

## 3. A Precise and Grand International Metropolis: Dadu's City and Architecture in the Yuan Dynasty

In 1271, Kublai founded the Yuan Dynasty as its first emperor, and renamed Jin Zhongdu to Yuan Dadu the next year. In 1279, the Yuan Dynasty conquered the Southern Song Dynasty and united China. Yuan Dadu was located in the northeastern suburb of Jin Zhongdu, initially built in the 4th Zhiyuan Year of the reign of Emperor Kublai (1267 A.D.), and the palace area was completed in 1274. Yuan Dadu was basically completed in the 13th Zhiyuan Year (1276 A.D.), and afterwards the construction continued. Yuan Dadu was another capital with a precise layout since Chang'an City of the Sui and Tang Dynasties. With precise layout, terrific scale and grand architecture, it was one of China's most famous capitals that had a very high status in the world's city history.

In layout, Yuan Dadu was the closest to the ideal put forward in *Kaogongji* (Study of Craftworks, a book completed in the Eastern Zhou Dynasty), with ideal order chased by the highest rulers of China's feudal society, i.e. square outer city wall, northern-southern axis of the whole city as the symmetry axis, eastern-western symmetry pattern, royal city and palace city at distinguished positions on the central axis, strictly horizontally and vertically orthogonal street grids, "Zu" (Ancestral Temple) built on the left and "She"(Altar of Land and Grain) on the right, and palaces with the plaza in front and the market at the back. The northern-southern axis was determined at first in the technical layout at the most important point on the eastern bank of Shichahai (today's Houmen Bridge in the north of Di'anmen) in the north of the North Sea. The place where Drum Tower stands was fixed as the geometric centre of the whole city, and the central platform was built there. Then the location of the southern wall was fixed according to the central platform's relation with the northern wall of Jin Zhongdu. The northern wall was fixed by taking the distance between the central platform and the southern wall as the radius. The western wall was fixed as per the western end of Jishuitan (water pool) in the west of Shichahai, enclosing all the water of Jishuitan inside the city. The eastern wall was fixed with the distance from the central platform to the western wall being the radius. However, the planned location of the eastern wall was found to be in a bottomland during the construction, so it was retreated a little, resulting that the distance of today's Beijing axis with 2nd Ring Eastern Road is a little shorter than that with 2nd Ring Western Road. The royal city of Yuan Dadu was located at the central of southern city axis. The royal city axis (the same as the axis of Beijing of the Ming and Qing Dynasties) was built with a palace city, or Danei, whose front gate was called Chongtianmen. Near Drum Tower in the north of the royal city was the most important market. The Imperial Ancestral Temple and the Altar of Land and Grain were built respectively on the left and the right outside the royal city and inside Qihuamen and Pingzemen, the western and the eastern city gates of the capital.

The Dadu City was rectangular in shape. The eastern and the western walls were a little longer than the southern and the northern ones, 6,700 meters from east to west and

7,600 meters from north to south. The City walls of Yuan Dadu were all built with rammed earth, and Dadu city had 11 gates (or "men" in Chinese), including Shunchengmen, Lizhengmen and Wenmingmen in the southern wall, Jiandemen and Anzhenmen in the northern wall, Guangximen, Chongrenmen and Qihuamen in the eastern wall, as well as Suqingmen, Heyimen and Pingzemen in the western wall. The names of the gates originated from *Book of Changes*. These gates don't exist now except that in 1970s when Xizhimen Tower of the Ming and Qing Dynasties experienced a demolition, the complete Heyimen of Yuan Dadu was found inside the city wall of Xizhimen Urn City (small city built near outside of the main city for defense purpose), which is today's only city gate relic of Yuan Dadu. Now, there remains a long city wall of Yuan Dadu near the intersection of Xueyuan Road and Xitucheng Road in the north of today's Xizhimen and extending to the north. There were broad avenues between city gates, "Nine Longitudes and Nine Latitudes" as *Kaogongji* put it. The square blocks divided by vertical and horizontal streets were called "Fang", in which were living or governmental districts. Dadu City had totally 50 such Fangs.

The successful layout of the water system was another characteristic of Yuan Dadu. Guo Shoujing, an outstanding scientist, served as the planner. He dredged Tonghui River, a canal in the east of Dadu, and also planned a new channel that originated from today's Baifu Spring in Changping District. Channels were built along the northern mountain foot in the direction of Dadu, caught springs from West Mount, shaped a lake in the north of the city, and then flowed into Tonghui River. The water system concentrated water from many sources, not only solving the water supply in Dadu, but also opening a transport canal that allowed southern materials to arrive in Dadu and solved material transport including the foodstuff.

The Yuan Dynasty regime held an attitude of tolerance and utilization towards various religions, and many large temples were built then, e.g. Huguo Temple, Miaoying Temple, Dongyue Temple etc. Though rebuilt by the offspring, their relics still remain some Yuan Dynasty style. Dongyue Temple maintains temple architecture characteristics of the Song, Jin and Yuan Dynasties. Lingyan Temple at Qiajiazhuang Village, Qingshui Town, Mentougou District in the suburbs of Beijing, still has architectural relics of the Yuan Dynasty. As for Tibetan Buddhism, most remaining buildings are upside-down alms bowl -style stupas, and the most famous one is the White Stupa in Miaoying Temple, Fuchengmennei Street, Xicheng District.

The form of quadrangle had matured in the Yuan Dynasty. There is no actual quadrangle of the Yuan Dynasty in Beijing now, and the only reference is the resident site excavated at Houyingfang in Xicheng District. According to the site, the quadrangle was surrounded by principal rooms and wing rooms in the Yuan Dynasty, shaping the most fundamental unit of urban architecture.

# 4. Model for Feudal Imperial Capitals of China: the City of Beijing and Its Architecture in the Ming and Qing Dynasties

After Zhu Di, Emperor Chengzu of Ming Dynasty ascended the throne in 1403, he decided to move the capital to Beijing. The capital construction started from the 5th year of the reign of Emperor Yongle of the Ming Dynasty (1407 A.D.) and completed in 1421. Beijing was then renamed Jingshi, but conventionally it is still called Beijing. The city was constructed based on the architectural experience of previous dynasties, and is a model of the national capital construction of Chinese feudalism empires, with the highest level in feudalism society in terms of design idea, planning and layout, and construction. Beijing is the largest and best-preserved city among Chinese ancient capitals. To date, basic pattern and many buildings established in the Ming Dynasty have been well preserved.

The city wall of the Ming Dynasty was based on the earth wall of the Yuan Dynasty. For the purpose of defense, the reserve land for city construction in the north of Yuan Dadu was given up, and the city wall was receded 2.5 kilometers southward, and the city wall in the south was expanded 800 meters southward from the today's Chang'an Street to the Xuanwumen and Chongwenmen. The city wall covered with bricks was in the shape of a rectangular. The city was 6,650 meters from east to west, and 5,350 meters from north to south. There were three gates in the southern wall, with Zhengyangmen in the middle, Chongwenmen in the east and Xuanwumen in the west; the eastern wall had two gates, namely, Dongzhimen and Chaoyangmen; the western wall had Xizhimen and Fuchengmen; the northern wall had Deshengmen and An'dingmen. High corner towers were built on the four corners of the city wall. The urban central axis of the Yuan Dynasty was reserved and it divided the city into two symmetric parts. The imperial city was located a little south to the center of inner city. It ran 2,500 meters from east to west and 2,750 meters from south to north, looking like an irregular square. In the center of the imperial city stood the most important palace complex, the Forbidden City, also known as the Palace Museum. The Forbidden City was 960 meters long from south to north, and 760 meters wide from east to west, with high city gates in all four sides, namely, the Meridian Gate to the south, the Xuanwu Gate to the north, the East Glorious Gate to the east and the West Glorious Gate to the west. Splendid corner towers were built on the four corners of the city wall. The principal structures were the front three halls and the rear two palaces, all standing along the central axis of the Beijing city. Behind the Forbidden City was the imperial garden in which high Longevity Hill (also called Zhenshan Hill or Jingshan Hill) was built as a background. All the roofs of the buildings in the palace area were covered with yellow glazed tiles. To reinforce the defense of Beijing city wall, an outer wall was constructed in the 32nd year of the reign of Emperor Jiajing of the Ming Dynasty (1553 A.D.). However, due to inadequate financial resource, the construction of the outer wall was discontinued after the southern outer wall was completed. Therefore, the wall and gate towers of the outer city were smaller than those of the inner city. The outer city wall in Beijing was 7,950 meters long from east to west, and 3,100 meters wide from north to south. The city wall in the south had three gates, namely, Yongdingmen in the middle, You'anmen in the west, and Zuo'anmen in the east; there were two gates in each of the eastern, western, and northern city walls, with Guangningmen and Xibianmen in the western city wall, and Guangqumen and Dongbianmen in the eastern city wall. On all of wall gates stood urn city, gate towers and arrow towers. The central axis of Beijing, totally 7.8 kilometers long, was the world's longest one of its kind from time immemorial. In addition, all important imperial buildings in the city were arranged on the axis or symmetrically along the axis. The central axis started from the Yongdingmen in the southern end of the outer city, extended northward, and ended at Drum Tower and Bell Tower. On the axis stood YongdDingmen, arrow tower of Qianmen, Qianmen, Damingmen, Chengtianmen, Duanmen, the Meridian Gate, the Gate of Imperial Zenith, the Hall of Imperial Zenith, Zhongji Hall, Jianji Hall, the Hall of Heavenly Purity, the Hall of Earthly Tranquility, Xuanwumen, Longevity Hill, and Bei'Anmen. On both sides of the central axis erected the Temple of Heaven, the Altar of the God of Agriculture, the Imperial Ancestral Temple, the Alter of Land and Grain, the Central and South Seas, and the North Sea. This arrangement made imperial buildings constitute the central part of the entire city, which derived from the ritual systems of feudalism society and the need of facilitating reign, but impeded the communications between the eastern and western regions in the city. The streets and alleys in inner city of Beijing basically followed the planning of Yuan Dadu, and were located on both sides of the imperial palace and government offices which stood on the central axis. The sketch map of these roads showed pane pattern formed by the intersection, which was the representative of the street layout of the ancient cities in China. On both sides of the main roads of various sizes were the areas for commercial and handcraft development, and around Hutongs or small lanes were quadrangles, which were the living quarters of ordinary people.

In addition, the tombs of the 13 emperors in the Ming Dynasty, commonly called the Ming Tombs or the Thirteen Tombs, were built at the foot of the Hill of Heavenly Longevity in Changping District of Beijing. The mausoleum area is surrounded by mountain range in the northern, eastern, and western sides, and faces the plain to the south. The tombs were built in the valley. In the central position lies Emperor Yongle's Tomb for Mingchengzu, the first emperor in the Ming Dynasty who moved the capital to Beijing. The other 12 tombs are distributed according to the terrain, with distances between each other ranging from 400 meters or 500 meters to 1,000 meters. On the mild slope in the front of the valley, and 6 kilometers away from the entrance of mausoleum area is the entrance to the mausoleum areas, with a huge stone archway erected there. In the north of the stone archway stands the gate to the tomb area. Along the street northward from this gate are the front stele pavilion, the Ornamental Column, the Dragon and Phoenix Gate, and 18 groups of stone statues of civilian officials and military officers, as well as stone animals such as elephant, horse and camel. Emperor Yongle's Tomb, built in the 22nd year of the reign of Yongle of the Ming Dynasty(1424 A.D.), is the largest one among the 13 mausoleums. Emperor Yongle's Tomb consists of the Tomb, the Square City, the Memorial Shrine, and the Ling'en Hall. The underground tombs of Ming Tombs built with huge stones are the "Subterranean Palace" which connects the rooms of the tombs. In 1956, archaeological workers unearthed Emperor Wanli's Tomb and opened it to visitors. The buildings on the ground in the Ming Tombs were mostly built in the Ming Dynasty, and a small part of tombs were restored in the Qing Dynasty.

The well preserved Ming Dynasty structure in Beijing is the world-famous Great Wall. The Ming Dynasty was established by overthrowing the sovereignty of the Yuan Dynasty which was then held by northern minority. The primary task facing the Ming Dynasty was to construct great walls to prevent the restoration of the former monarchy. Since Yuan Dadu was occupied in the first year of the reign of Emperor Hongwu of the Ming Dynasty (1368 A.D.), the passes, bodies of great wall, and beacon towers started to be built. Afterwards, nearly every emperor of the Ming Dynasty attached importance to the construction and repair of the Great Wall. The main bodies, namely city walls, were completed around 1600. The city walls, which were built mostly on the watershed lines of the mountain chains, has the following forms according to the regional characteristics: strip stone wall which was rammed with soil and concrete, block stone wall, earth-rammed wall, brick wall etc. In all strategic places where the Great Wall put through, passes of military significance with solid defense facilities were established. Important passes formed towns of military importance, for example, Juyongguan Pass in Beijing was the most important passage through which the forces moved into Beijing.

The preserved buildings built in the Ming Dynasty are the Taisui Hall in the Altar of the God of Agriculture, the Zhihua Temple, the Office of Divine Music in the Temple of Heaven etc.

In the 1st year of the reign of Emperor Shunzhi of the Qing Dynasty (1644 A.D.), the national capital was moved to Beijing. The imperial city of the Ming Dynasty remained in use, with the four gates unchanged. However, some alternations were made on the basis of the imperial city of the Ming Dynasty, i.e. Bei'anmen was renamed Di'anmen, Chengtianmen renamed Tian'anmen, Damingmen renamed Daqingmen, and the Longevity Hill was renamed Jingshan Hill (commonly called Coal Hill). In the 15th year of the reign of Emperor Qianlong(1750 A.D.), on Jingshan Hill were built five pavilions, among which the largest is Wanchun Pavilion lying on the central axis of the entire city, to the west are Guanmiao Pavilion and Zhoushang Pavilion, while to the east are Jifang Pavilion and Fulan Pavilion. During the Qing Dynasty, most of the palaces built during the Ming Dynasty in the North and the Central South Seas were dismantled or renovated, showing greater change from the former palaces. The White Stupa on Qionghua Island was built in the 8th year of Emperor Shunzhi (1651 A.D.), and Yong'an Temple was constructed on the site of Guanghan Palace in front of the stupa.

The Forbidden City of the Qing Dynasty maintained the architectural pattern inherited from the Ming Dynasty. However, most of the architecture built during the Ming Dynasty had been destroyed by wars at the end of the Ming Dynasty. The restoration of these architecture started in the first year of the reign of Emperor Shunzhi, and was almost completed in the 25th year of the reign of Emperor Kangxi (1686 A.D.). Afterwards, with support of ample treasury funds, these buildings went through constant construction and expansion, and developed into a large scale during the reign of Emperor Jiaqing. Almost all the Ming Dynasty architecture built in the Forbidden City had gone through renovation, and there were no remainder of the Ming Dynasty left. The northern gate, which was called Xuanwumen during the Ming Dynasty, was renamed Shenwumen (the Gate of Martial Spirit) in the Qing Dynasty, the other three gates retained the former names, i.e. the Meridian Gate, the East Glorious Gate and the West Glorious Gate. The Forbidden City in the Qing Dynasty still used the style of the Ming Dynasty, and the main body constructions remained such that the front area known as the Outer Court was used as a place where the emperors handled state affairs while the Inner Palace were used as living quarters of imperial family.

The Qing governors preferred to establish temporary palaces in the mountain and water areas in the suburbs of Beijing, thus royal gardens called "Three Hills and Five Gardens" formed in Xishan (Western Hills) of Beijing. These gardens were all located in the western region of the Beijing suburbs, namely, Jingyi Garden in Fragrant Hill, Jingming Garden in Yuquan Hill, Qing Yi Garden in Longevity Hill, Yuanmingyuan Garden, and Changchun Garden. Of the five gardens, Yuanmingyuan Garden boasted of the largest scale. The construction of the garden started in the 48th year of the reign of Emperor Kangxi of the Qing Dynasty (1709 A.D.), and later was expanded by Emperor Yongzheng into a royal villa, and was enlarged again by Emperor Qianlong and was completed in the 9th year of the reign of Emperor Qianlong (1744 A.D.). The construction of Changchun Garden and Yichun Garden beside the Yuanmingyuan Garden was completed in the 37th year of the reign of Emperor Qianlong(1772 A.D.).

In addition, to unite the Mongolian and Tibetan people, the Qing government built a large amount of Lama temples, among which the well preserved temples are Songzhu Temple, Fuyou Temple, Miaoying Temple, Huguo Temple, Xihuang Temple, Longyu Temple and Yonghegong Lama Temple etc.

Furthermore, the Qing government prescribed that all imperial clans built official residences in the capital of Beijing, instead of granting them fiefdoms. The private residences of imperial families also constitute a unique architectural element. The structural scale and patterns of royal families in the Qing Dynasty, with superb scale, solemn layout, multiple courtyards, and magnificent color painting, decoration with colored glaze, are only next to imperial architecture, and have a special important status in the traditional architectures in Beijing.

The last is the quadrangles of various sizes in Beijing. According to the records of *The Map of the Capital City of Qianlong Period* which was drew in the 15th year of the reign of Emperor Qianglong of the Qing Dynasty, there were more than 40,000 quadrangles with complete pattern inside the Beijing city at that time, and quadrangles were the major form of residence in Beijing. In the Qing Dynasty, the fixed architecture pattern had been formed which enclose the lined houses into quadrangles.

Through nearly fifty years of development since the founding of the People's Republic of China, Beijing has become a modernized city where ancient and modern architecture are interweaved, making the protection and development to be the eternal task for human beings.

Hou Zhaonian
February,2007

宫
殿

PALACES

# 北京的宫殿

中国历代统治者都要大规模地建造宫殿，自商周至清代，宫苑繁盛，如绵延上百里的秦代阿房宫，宏丽壮观的汉代长乐宫、未央宫，唐代大明宫等。汉代时形成了一套理论，即《周礼·考工记》规定的"匠人营国，方九里，旁三门，国中九经九纬，经涂九轨，面朝后市，左祖右社……"①。这套原则一直指导着以后的历代王朝皇宫的营造。北京作为封建社会中后期的政治中心，从辽代到清代都建造了规模宏大的宫殿。虽然这一时期的宫殿规模较前代占地面积少，但是宫殿集中，布局严谨，利用各种建筑手法营造出巍峨、严肃、神圣的气氛，并且留下了举世闻名的明清两代皇宫——紫禁城。

## 一、北京宫殿建筑的发展史

### 1.辽金时期北京宫殿的初创

商周时期北京地属古燕国和古蓟国，那时的王宫建筑早已湮灭，史书上也只有"蓟丘"这一称谓，也只能让人们遐想这一当年诸侯国宫殿的形象了。以后各代都只是中央王朝的一个州郡，也没有什么大规模的宫殿建造，直到辽金时期，北京作为一个独立政权的首都才开始大规模建造宫苑。

辽代虽然大量保留了游牧民族的习惯，皇帝一年当中多在行宫中度过，但是在其南京(即北京)"利用燕蓟等地的工匠建造官署"。根据《辽史》记载可知辽南京城分为大城、皇城和宫城②，皇城位于大城的西南角，皇城内又分为东部的宣和宫和南部的大内两部分。大内(即宫城区)为主要宫殿区，相当于明清的紫禁城，不是位于皇城的中心而是位于皇城南部，大内东面为永平馆，为接待来访、朝贡使者的地方。此外，据研究宫殿西侧为瑶池宫苑区。宫苑规模较大，瑶池中有小岛瑶屿，上有瑶池殿，池旁建有皇亲宅邸。从总体上看，这时北京地区的宫殿还属于初创期，宫殿不是位于整个城市的中心，规模也不是很大。

金代辽后，1149年金帝完颜亮继位后，准备将都城迁到燕京(即今北京)，天德三年(1151年)开始扩建燕京城和

宫殿。金中都的宫殿是按照宋代东京汴梁的宫殿，并适当展拓规模而成；另外，还拆卸了东京宫殿的一部分装修、装饰类的建筑构件运往中都。两年后宫殿建成，金朝正式迁都，改名中都。中都在辽南京城故址上兴建并扩展，平面近似方形，同样分为大城、皇城、宫城三部分。皇城同样没有位于整个都城的中心，而是位于都城中部偏西。皇城南部东侧为太庙，西侧为内省，中间是千步廊，千步廊中间夹御道；皇城西部为同乐园，是苑囿建筑区；同乐园北为北苑，内有湖泊、景明宫等建筑；皇城东部为东苑和内省，是宫苑区和为皇宫服务的机构所在地。

宫城位于皇城中央偏东(据后来学者考察，金代宫殿位于今广安门一带)，宫城周围九里三十步，"殿凡九重，殿三十有六，楼阁倍之"③，遵循了皇帝居九重宫阙说法。宫

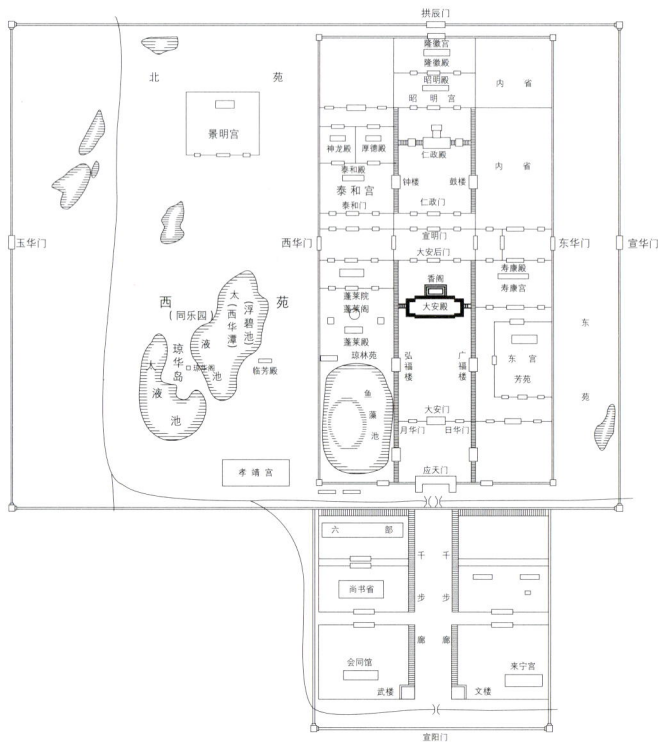

金中都皇城宫城总体布局示意图
转引自郭黛姮主编《中国古代建筑史》第三卷

城正中为一条中轴线，贯穿南北，最重要的宫殿都布置在中轴线上，中轴线向南延伸，通过皇城南门宣阳门及都城南门丰宜门，向北延伸通过皇城北面的拱辰门直达都城北门通会门。中轴线分为前部的外朝区和后部的内廷区两部分，遵循了"前朝后寝"的布局原则。沿中轴线左右分别为东路和西路。东路为太子居住的东宫和太后居住的寿康宫及内省。西路有御花园，如琼林苑、蓬莱院，以及妃嫔居住的十六位(即寝宫)。此外，应该提出的是西路南部是辽南京的瑶池和瑶屿，金称为鱼藻池。

综观金中都宫殿，虽然是少数民族政权所建，而且很多殿宇的装饰也体现了其民族特色，如一些殿宇和院子内铺毡毯、动物皮毛等，但是由于金中都的宫殿大体是模仿宋汴梁的宫殿建造，又有宋朝技术人员的参与建设和设计，设计思想、建筑布局和建筑样式都继承了历代宫殿的传统，所以金中都宫殿可以说是12—13世纪时期中国宫殿建筑的代表作。

## 2.元大都宫殿

元太祖十年(1215年)置燕京路总管大兴府，世祖至元元年(1264年)，元世祖忽必烈着手元大都的兴建，至元四年(1267年)，在辽金故城东北创建新城，派汉族儒士刘秉忠负责规划建设，其中宫殿是大都的主要建筑。建成后的宫殿分为皇城和宫城。皇城正门承天门，外有石桥和棂星门，再往南，御街两侧建长廊，称"千步廊"，直抵都城的正门丽正门，皇城东西两侧建有太庙和社稷坛。宫殿集中在皇城以内的宫城，宫城"周回九里三十步，东西四百八十步，南北六百十五步，高三十五尺，分为六门，正南曰崇天、左曰星拱，右曰云从，东曰东华，西曰西华，北曰厚载"④。宫城包括三组宫殿和太液池、御苑等建筑。宫城位于全城中轴线的南端，又称大内。宫城之西是太液池，池西侧的南部是太后居住的西御苑，北部是太子居住的兴圣宫，宫城以北是御苑，宫城四角建有角楼。

元大都宫殿承上启下。首先，宫城内以大明殿、延春阁为主的两组建筑都建在大都城的南北中轴线上，配殿沿

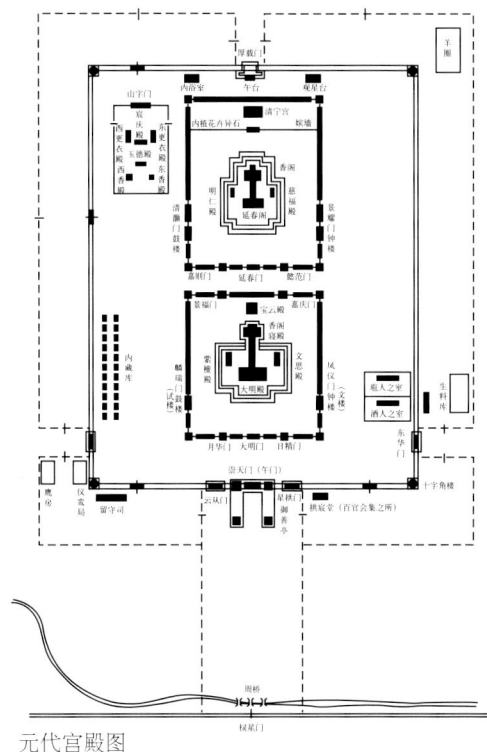

元代宫殿图

轴线左右对称排列，这既是继承了宋、金建筑的布局形式，也开启了明清北京城和紫禁城的中轴线。其次，元大都宫殿还继承了宋代宫殿采用的四角设角楼、四面设门、崇天门前设周桥(礼仪性的桥)和千步廊、大明殿和延春阁使用"工"字殿等，这些做法明清紫禁城也都继承了下来。再就是大都宫殿利用了金中都的太液池离宫旧址环水布置宫殿，如大内和隆福、兴圣二宫分散设置于太液池东西两侧。究其原因虽然是因为蒙古游牧民族逐水草而居的生活习俗，便于洗马、饮马，客观上也为明清两代进一步开发御苑，形成后来的南海、中海(后来南海和中海合称中南海)和北海奠定了基础。大都宫殿在建筑内容上也体现了蒙古民族的特色，如蒙古人喜欢饮酒，于是东华门北建有庖人之室、酒人之室，供殿宴执事人员居住。再如建有鹰房、羊圈，也是其习俗的体现。宫殿室内装修上也体现了蒙古族特色，如正殿内设两个宝座，帝、后并列而坐，皇后也参与朝会，诸王、文武官员都有坐床(其他殿宇也多如

是)。喜欢用动物毛皮作装饰，帷幔、壁幛、地衣等都用动物毛皮。再有就是大酒瓮是元代宫殿内必不可少的陈设品。总之，元代大都宫殿是中国封建社会宫殿的又一座代表作。

### 3.辉煌定型的明清北京宫殿

明代建立后，初定都南京，实行诸王"分封制"，诸王各霸一方，北京为燕王朱棣的封地。朱棣通过"靖难之役"夺取皇位后，出于政治、军事等各方面的考虑决定迁都北京，并开始修建北京宫殿。

明代北京宫殿的修建分为两个时期。第一，永乐朝创建。朱棣为燕王时的府邸利用了元朝故宫，称帝之初也曾想要建北京宫殿。但是由于当时战争初息，百业待兴，所以只是在元代故宫的基础上建奉天三殿以备巡幸时使用，没有新建宫殿。永乐十四年(1416年)，朱棣准备巡幸北京，故而商议营建宫殿。至永乐十五年(1417年)开始改建皇城，选址在全城中心稍偏南，元代旧宫东一里左右，皇城内主要布置庙社、宫苑、寺观、衙署、仓库等建筑。皇城中央即为紫禁城，宫殿完全仿照南京宫殿的布局，只是规模更加宏伟，至永乐十八年(1420年)建成。建成后的宫殿宫城周六里一十六步，亦曰紫禁城。门八：正南第一重曰承天，第二重曰端门，第三重曰午门，东曰东华，西曰西华，北曰玄武。紫禁城内宫殿分为外朝区和内寝区两部分。外朝区自午门开始，午门正北为奉天门，是皇帝平日上朝的地方，过奉天门即外朝正殿奉天殿(即金銮殿)，其后为华盖殿和谨身殿；内寝区以乾清、坤宁两宫为主要建筑。这时北京的宫殿位置及四至范围已经基本上形成。

第二，后代的重、扩建。紫禁城宫殿建成的第二年，也就是永乐十九年(1421年)，前三殿被火焚毁。永乐二十年(1422年)，乾清宫也遭火灾。殿宇焚毁之后，朱棣认为这是"天意"，没有再修。直至英宗正统五年(1440年)才复建。此外，宣德年间(1426—1435年)由于居民喧嚣声太大，再次扩建了紫禁城，将紫禁城东门(即东华门)跨过元故宫的护城河向东扩展了一部分，使得紫禁城范围进一步

扩大。景泰六年(1455年)，增建了御花房。天顺三年(1459年)，经营西苑，新增行殿三处。至此，北京宫殿及御苑基本完备。至嘉靖朝(1522—1566年)又开始了对故宫的新一轮扩建，改扩建了20余处，使北京宫殿进入了明朝的极盛期。其中最大的是"嘉靖三十六年(1557年)丁巳四月十三日，奉天等殿火灾，是日申刻雷雨大作，戌刻火光骤起，由正殿延烧，午门楼廊俱尽，次日辰刻始熄。三十七年七月大朝门等工成"⑤。嘉靖四十一年(1562年)三大殿改奉天殿为皇极殿，殿门曰皇极门，华盖殿为中极殿，谨身殿为建极殿。正德九年(1514年)，乾清宫和坤宁宫遭火灾，万历二十四年(1596年)乾清宫再次遭火灾，次年重建。万历二十五年(1597年)，三大殿遭火灾，天启五年至七年(1625—1627年)重建，此次重建耗费银两5957519余两，由此可见宫殿修缮耗资之巨大。至此，经过历代的修、扩建，明朝的皇宫恢宏大气。

此时的宫殿与明前期相比出现两点变化：一方面，生活休闲气息更浓。前期宫殿政治色彩浓烈，生活上强调节俭，但是随着后代追求安逸生活的需要，宫殿更加强调舒适享受，宫殿开始园林化。如朱瞻基就把其祖父朱棣骑射场地东苑改成斋居别馆，英宗改建成正式离宫，到嘉靖、万历时，随着南内的出现，西苑的扩建，万岁山(今景山)的开辟，慈宁宫花园、坤宁宫花园的兴建，北京宫殿的内容和性质已经大异于前期。另一方面，由于帝后信佛崇道，北京宫殿内增建了很多宗教场所和佛殿。另外，明代的宫殿使用了大量名贵的木料，如楠木、檀香木等，而到了清代这些名贵的木料几乎已经被消耗殆尽，尤其是大块木料，更加稀缺，甚至同样大的一块木料就要用同样大的一块黄金换取。

满清入关之后，宫殿已经在此之前被李自成焚毁殆尽，清朝并没有重新选址兴建宫殿，而是在明代基础上修缮、复建，制度规模一仍明旧，改变殊少。顺治二年(1645年)修缮各处殿宇，定前三殿名太和、中和、保和，后宫名则少改动，次年完工。顺治十二年(1655年)重修内宫。"康熙八年(1669年)，敕建太和殿，南北五楹，东西广十一

楹。十八年（1679年）太和殿灾。二十九年（1690年）重修三殿，三十六年（1697年）工成。至此大内修建至清初已告一段落，诸宫殿皆经重修或重建，然无一非前明之旧规也"⑥。乾隆三十年（1765年）重修三殿。自此以后三大殿就没有再改建，所以今天的太和、中和、保和三大殿即是当时修复之面目。乾隆三十九年（1774年）敕建文渊阁于文华殿后，用以珍藏钦定《四库全书》。嘉庆二年（1797年），乾清宫和交泰殿发生火灾，是年重修，次年完工，以后未大修葺。

纵观大内沿革，一切巨规宏模，无一不沿自明朝。然其修筑之宏，抑又不逮。康熙二十九年（1690年），诸臣等复奏云："查故明宫殿楼亭门名共七百八十六座，今以本朝宫殿数目较之，不及前明十分之三。考故明各宫殿九层，基址墙垣，俱用临清砖，木料俱用楠木；今禁内修造房屋出于断不可已，凡一切基址墙垣，俱用寻常砖料，木植皆松木而已。"两代营建，优劣之势，于此可见。综观清代，大工可数，火灾亦少，唯满人颇能保守，故能汇为大观，保存至今。宫内还有一座特殊的建筑，即御花园，位于宫城中路最后部。始建于永乐年间，景泰、万历迭予增筑，有清一代，革易极少。其间奇石罗布，佳木葱郁，古柏老藤，皆明代旧物。御花园与宁寿宫的乾隆花园及慈宁宫花园，并称胜境。

另外，清代紫禁城虽然基本格局形成于明代，但是清代相较于明代也有若干变动和改动。第一，局部建筑的改、扩建：外朝保和殿的位置向后移动位于"工"字台基的后沿了，而明代的建极殿后面还有一座后门——云台门。乾西五所改建为重华宫。明朝宫城内廷以乾清宫、坤宁宫为紫微正中，左右各有东西六宫以为辅翼。在东西六宫以北，各有五所供皇子、皇孙居住的次要宫室称为乾东房五所、乾西房五所，平面规划上在乾清宫左右。明末内廷宫室焚毁，东西六宫是在顺治十二年（1655年）及康熙二十二年（1683年）分两次陆续复建完成，至于乾东西五所的复建约在康熙到乾隆年间，如今日的规制，乾西五所改建为重华宫、漱芳斋、建福宫等宫室建筑，相较于明代变化较大。改建后的乾西五所成为一座宴赏、集会及园林性质的

宫室，平面布置上较为灵活。可惜这么一座精美的园林化宫室在民国十二年（1923年）全部被烧毁，仅存一座大假山。其次是宁寿宫的改建。宁寿宫位于内廷外东路，占据了宫城东北角，东西宽120米，南北长395米，占地4.74公顷，原来为明代的东裕库与仁寿殿旧址，北部为居住老年宫妃的哕鸾宫、喈凤宫旧址。第二，建筑的使用性质的变化：后三宫在使用性质上发生了重大变化，尤其是坤宁宫。明代坤宁宫为皇后的日常居住处，清顺治十二年（1655年）复建坤宁宫时将其改为祀神、皇帝大婚的处所，而且由于皇帝大婚也只是在此居住两日，就搬到乾清宫和养心殿，所以这里实际上成了神堂。清代康熙朝以乾清宫为内廷理事之所，所以御门听政由明代的奉天门（皇极门，清代称太和门）改为乾清门进行，从康熙至咸丰朝前后六朝都在此。雍正朝将内廷中心改为养心殿，所以养心殿也相应地作了改造，并且以后都以此为清宫内的政治、起居中心，直到清亡都未改。乾东五所在嘉庆十四年（1809年）后改为如意馆、寿药房、敬事房、四执库、古董房等。由此我们可以看出，清代的紫禁城虽然继承了明代的格局和基址，但是出于满族的民族习惯和居住要求等各种原因，局部建筑上也稍有变化，体现了清代紫禁城的特色。

1911年，孙中山领导的辛亥革命推翻了清王朝，根据协定清皇室仍居后宫。1914年，在前三殿成立古物陈列馆，故宫开始作为博物馆。1924年冯玉祥将清逊帝溥仪驱逐出皇宫。抗日战争期间，为了保护珍贵文物，大量文物南迁。新中国成立后，故宫仍然作为博物馆直至今日。

## 二、北京宫殿的建筑、艺术成就

### 1.古今比较

纵观中国古今，北京的宫殿在占地规模上比早期和中期的宫殿小，如秦朝咸阳宫殿以及阿房宫蜿蜒上百里，汉代的未央宫、长乐两宫，分别为4.6平方公里和6.6平方公里；唐长安城的大明宫占地面积3.3平方公里，占地面积都很广阔，而北京紫禁城占地面积仅0.73平方公里。但是，早期宫殿更多地依托自然环境的衬托，宫内布置多组小的

宫殿，而这些小宫殿自成一区，相互之间联系不是很密切，富于园林气息，宫苑的性质更突出。而中后期的北京宫殿，占地面积上虽然比中前期小很多，宫殿基本上集中在皇城以内，但是这一时期宫殿的布局更加严谨、紧凑而合理，能够在有限的范围内充分利用空间，利用建筑手法烘托环境，能够在有限的空间内通过高低错落、抑扬顿挫等建筑手段，创造出不同的建筑意境来，从而达到建筑所想要表现的功能需求，充分体现建筑本身的感染力，是千百年来中国建筑的集大成的代表作。

## 2.中外比较

中国古建筑自成一体，马可·波罗在其著作《马可·波

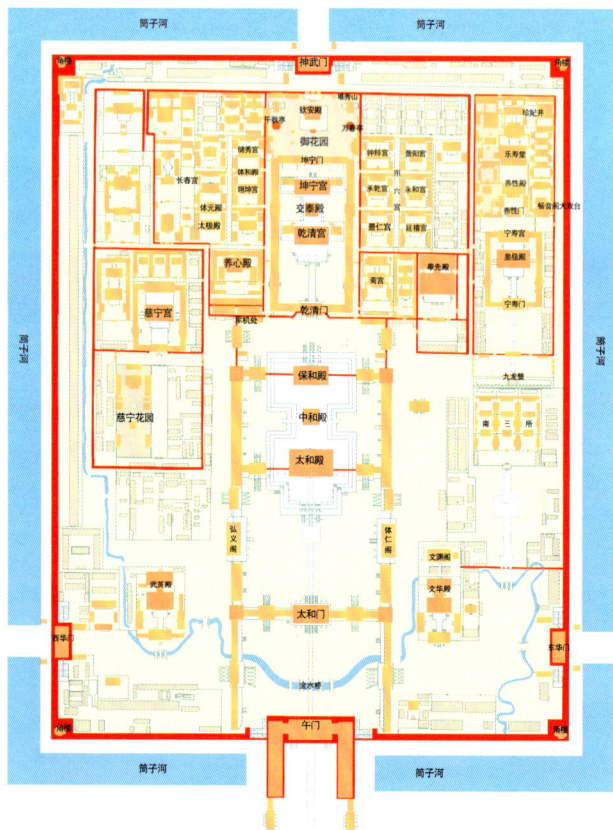

故宫示意图

罗游记》中对大都城的宫殿赞叹不已，引起了西方世界的极大震动。另外，现存的作为中国古建筑代表作的北京"紫禁城"是世界现存最大的、历史最悠久的、保存最完整的一座宫殿建筑群，与世界著名的皇宫建筑相比，其占地面积是法国卢浮宫的4倍，是英国白金汉宫的10倍，是俄罗斯冬宫的9倍，是号称欧洲最大的宫城——克里姆林宫的2倍还多，日本东京的皇宫建筑面积也只有紫禁城的1/3。无论在建筑技术上，还是建筑艺术上北京的宫殿都取得了很高的成就，而且其影响直接波及日本、朝鲜等国家的宫殿建筑。

## 3.璀璨明珠紫禁城

保存至今的明清宫殿紫禁城从规划布局上承袭了历代帝王所遵循的礼制制度、周王城制度，进一步规范化了唐宋以来宫殿营造的模数制，采用了中国传统文化的风水学说，与历代皇宫一脉相承；在建筑意境方面，利用建筑群烘托皇帝的威严与神圣，达到了登峰造极的程度；在建筑技术上更加成熟、建筑设计与施工高度标准化，是中国封建社会后期大型建筑群的巅峰之作。正如梁思成先生在其著作《中国古代建筑史》中所说："其规模之宏伟，已世无与伦比矣"[7]。

首先，紫禁城将中国传统的轴线对称发挥到极致。紫禁城的主要宫殿均建在一条南北向长1.6公里的中轴线上，用连续的、对称的封闭空间和一系列的廊庑、门殿，形成逐步展开的建筑序列，来衬托出三大殿的宏伟、威严和神圣。此外，这条中轴线与整个城市中轴线重合，自永定门到大清门，向北经宫殿轴线至鼓楼、钟楼，贯穿了整个城市，使得宫城轴线南北延长，更增强了宫殿的轴线气势，成为世界上最长的一条城市轴线。

其次，色彩的成功运用。明清紫禁城全部采用黄色琉璃瓦，红色墙体，由于其又位于城市中心，从城市上空鸟瞰，这片高大而又金碧辉煌的建筑被城市周围低矮的灰色建筑物包围着，形成鲜明的对比，显得非常醒目。这种色彩对比的方法使宫殿达到了宏大和堂皇的效果。

再次，建筑意境的成功运用。紫禁城成功地运用建筑

物营造出了各种氛围，使进入其中的人们产生种种的心理变化，以表现出皇宫既威严肃穆，又宏大富丽的效果，从而达到其所想达到的皇权至高无上和皇帝的神圣感。如紫禁城以一座低矮的大清门(明代称大明门)作为开端，给人的第一感觉是平淡，大清门向北紧接一个长500余米的低矮长廊——千步廊组成的狭长的前院，造成人的压抑、低沉和漫长的感觉，把人的情绪压到低点。而长廊的尽头却突然开阔起来，前面是一个深300米的横向空间，巍峨的皇城正门——天安门矗立在空间的北段正中，门前配以汉白玉金水桥和华表，凸现了皇城的宏大，从而形成了第一个建筑高潮，而此时人的心理也会随着宏伟的建筑物突然高涨，产生对皇帝居住之所——皇宫的崇敬之情。越是走近天安门，就会越来越感觉天安门的高大。进入天安门，前面是一个狭小庭院，庭院北端是形式和天安门相同的端门，两门连续的覆压更显示出皇城的威严气氛。过端门再次进入一个深300余米的狭长院落，两旁为林荫道和低矮的房子，高大的建筑物和幽暗的林荫道使人感觉威严和压抑。行至紫禁城正门午门前，高大的午门平面呈倒"凹"字形怀抱着一个较小的空间，连门也是棱角分明的方形，在经过起初到此地的漫长行进后，由于体力和心理的疲惫，

置身其中，森严、压抑的感觉更加浓重，从而把皇宫的威严气氛推到又一个高潮。过午门是宽度达200余米的太和门庭院，至此建筑空间豁然开朗，远端的门洞也变成圆形，庭院中间流淌的金水河和桥上洁白轻盈的汉白玉金水桥，使人从极度压抑中再次高扬，再次给人心理上制造了一个很强烈的对比，也为下一个高潮作了铺垫。过太和门，紧接着又是一个面积达4公顷多近乎方形的更大的广场式庭院，太和殿及广场的气势扑面而来。太和殿建在面前正中三重汉白玉高台上，台上汉白玉柱头栏板回环排列，太和殿采用重檐，四周10余座门、楼、廊庑环列拱卫，远望宛若天宫，更凸显了皇宫的神圣，建筑达到了全局的最高潮。步行向前逐渐接近太和殿时，会由开始的大的感觉逐渐变化为高，继而三大殿的精雕细琢、富丽堂皇让你感到目不暇接。过三大殿后至乾清宫，宏大的气势逐渐退去，华贵的感觉成为主导，台基改为色彩华丽的琉璃栏墙，室内装修更是使用名贵的檀木等木料，人置身其中便感觉到皇帝的奢华和富有。至御花园，高树蔽天、假山掩映，清风油然而生，建筑布置也开始灵活多样，给人一种清凉和如释重负的感觉。宫殿后面以御苑内高高的景山和五亭作景的收束，使皇宫有了举目远眺的目标。

注释：
① 郑玄注：《周礼·考工记》。
② 中国的都城一般分为城、皇城和宫城，皇城以内即为禁地，一般百姓不能进入，属于宫苑、庙社、衙署区。皇城最主要的部分为宫城，即宫殿区。明代嘉靖时在大城南面加砌一道外城，清代时将皇城墙拆除。
③ 宋·宇文懋昭：《大金国志》第三十三卷，燕京制度、南沙席氏刻，乾隆朝版本。
④⑤⑥ 清·孙承泽：《春明梦余录》卷六，宫阙，古香斋本 光绪九年孟春刻成，"元大都宫殿考"。
⑦ 梁思成：《中国古代建筑史》，245页，北京：百花文艺出版社，1998年8月第5次印刷。

# Palaces in Beijing

Chinese rulers of the past dynasties all built their palaces in a large scale. Beijing, as the political center in the middle and late of feudal society, saw large scale palaces built from the Liao Dynasty to the Qing Dynasty. Though the palaces in such a period occupied a smaller acreage than those in the past dynasties, they were concentrated, had a precise layout, and created a lofty, solemn and holy atmosphere with various architecture designs. A world's famous palace of them is the Forbidden City, which had been the imperial palaces of the Ming and Qing Dynasties.

## 1. Development History of Palace Architecture in Beijing

### (1) The origination of Beijing palaces of the Liao and Jin Dynasties

According to *History of Liao*, Nanjing City included the great city, the royal city and the palace city. The royal city was at the southwestern corner of the great city, with Xuanhe Palace in the eastern part and Danei (the palace city area) in the south. The palaces in Beijing area were still in the initial period, not at the central of the whole city and in a scale not so large.

In the Jin Dynasty, Wanyan Liang, King of Jin, succeeded to the throne and made preparations for migrating the capital city to Yanjing (today's Beijing) in 1149. The third Tiande Year (1151 A.D.) saw the beginning of extension of Yanjing City and palaces. Jin Zhongdu was built at the site of Nanjing City of Liao Dynasty and extended, with a similarly square plane and divided into the great city, the royal city and the palace city. The royal city was not at the central of the whole capital city, but a little west. The Imperial Ancestral Temple and Office of Imperial Affairs were at the eastern and western sides respectively in the southern part of the royal city. In the middle was the thousand-step corridor, the royal road was inside it. In the west of the royal city was Tongle Garden, a garden architecture area, in the north of which was

Beiyuan (North Garden) including lakes, Jingming Palace and other buildings. In the eastern part of the royal city were Dongyuan (East Garden), including the Imperial Garden area and the Offices of Imperial Affairs, both served for the imperial palaces.

The palace city was a little east of the central of the royal city, in accordance with the principle that emperors live in 9 rings of palace. The central axis crossed the middle of the palace city from the south to the north, and key palaces were all on it. It extended southwards through Xuanyangmen and Fengyimen (the southern gates of the royal city and the capital city respectively) and northwards through Gongchenmen and Tonghuimen (the northern gates of the royal city and the capital city respectively). The central axis went through 2 parts: the outer palace and the inner court in front and at the back respectively, in accordance with the layout principle that the court is in front and living palaces at the back. The palaces in Jin Zhongdu were built mainly by imitating those in Bianliang of the Song Dynasty, with technical people from the Song Dynasty participating in the design and construction. The mind of design, architectural layout and patterns inherited the tradition in palaces of the past dynasties. So the palaces in Jin Zhongdu were a tour de force of Chinese palace architecture in the 12th and the 13th centuries.

### (2) Palaces in Yuan Dadu

The 4th Zhiyuan Year of the reign of Emperor Shizu of the Yuan Dynasty (1267 A.D.) saw a new city built in the northeast of the former city of Liao and Jin Dynasties, and Liu Bingzhong, a scholar of Han Nationality, was appointed to plan the construction. Palaces were the main architecture in Dadu. The palaces completed were divided into the royal city and the palace city. The main gate of the royal city was Chengtianmen, out of which were a stone bridge and Lingxingmen. Southwards, the imperial street, flanked by corridors known as thousand-step corridor, extended to Lizhengmen, the main gate of the capital city. On the western and eastern sides of the royal city there respectively built the Imperial Ancestral Temple and Altar to the Gods of Land and Grain. The palaces concentrated in the palace city, which was within the royal city. The palace city, including 3 groups of palaces, Taiye Pool and Imperial Garden, was on the southern end of the central axis of the whole city, also known as Dannei. In the western part of the palace city was Taiye Pool, in the southwest of which was West Imperial Garden where lived the queen mother, and in the northwest of which was Xingsheng Palace where lived the crown prince. The Imperial Garden was in the northern part of the palace city, which had a corner tower on each of the four corners.

The palaces in Yuan Dadu served as a milestone of palace architecture in Beijing. First, 2 groups of architecture within the palace city, Daming Hall and Yanchun Pavilion as the representatives, were built on the south-north central axis of Dadu City. The side palaces were arranged symmetrically along the axis, which not only inherited the architecture layout of the Song and Jin Dynasties but also inspired the central axis in Beijing city and the Forbidden City of the Ming and Qing Dynasties. The palaces in Yuan Dadu also inherited the palace layout of the Song Dynasty that there set corner towers at 4 corners, gates in 4 directions, Zhou Bridge (a bridge for rites) and thousand-step corridor in front of Chongtianmen, as well as Daming Hall and Yanchun Pavilion forming "工" shape etc. The Forbidden City of the Ming and Qing Dynasties inherited all these designs. The palaces in Dadu also featured the characteristics of Mongolians in architecture content. To sum up, the palaces in Dadu of the Yuan Dynasty represented another tour de force among those in Chinese feudal society.

## (3) Brilliance and finalization of palaces of the Ming and Qing Dynasties in Beijing

Zhu Di, Emperor Yongle of the Ming Dynasty who seized the throne from his nephew, migrated the capital to Beijing and built palaces in there. The construction of palaces of the Ming Dynasty in Beijing can be divided

into 2 periods. The first was the founding period of Emperor Yongle. The royal city started to be rebuilt in the 15th year of the reign of Emperor Yongle, and the site was selected a little south of the city center. In the royal city were mainly Imperial Ancestral Temple and Altar to the gods of Land and Grain, palaces, gardens, temples, administrations, warehouses etc. In the middle of the royal city was the Forbidden City, and the palaces completely followed the palace layout in Nanjing but featured a grander scale. The 3 major palaces were completed in the 18th year of the reign of Emperor Yongle. The palaces in the Forbidden City can be divided into 2 parts: the outer palace and the inner court. The outer palace area began from Wumen, in the north of which was Fengtianmen, where the emperor met officers and handled national affairs. Inside Fengtianmen was Fengtian Hall, the main hall of the outer palace area of the Ming Dynasty. Behind Fengtian Hall were Huagai Hall and Jinshen Hall. The inner court area had the Hall of Heavenly Purity and the Hall of Earthly Tranquility as the major architecture. The palace locations and 4 boundaries of today's Beijing had shaped basically.

The front 3 halls were burnt down in the 19th year of the reign of Emperor Yongle (1421 A.D.), and the Hall of Heavenly Purity also suffered a fire the next year. They weren't rebuilt until the 5th year of the reign of Emperor Zhengtong (1440 A.D.). The Forbidden City was extended again during the reign of Emperor Xuande. The 6th year of the reign of Emperor Jingtai (1455 A.D.) saw an imperial greenhouse added. Xiyuan (West Garden) was developed and 3 temporary imperial palaces added in the 3rd year of the reign of Emperor Tianshun. Then, the palaces and Imperial Gardens in Beijing were completed basically. Emperor Jiajing began a new round of extension of the Imperial Palace, involving more than 20 rebuilt or extended places, which brought the palaces in Beijing to the culmination of the Ming Dynasty.

The Qing Dynasty didn't select a new site to build

palaces, but repaired and rebuilt based on those of the Ming Dynasty, and the principle and scale inherited the Ming Dynasty with few changes. The palaces and halls were repaired in the 2nd year of the reign of Emperor Shunzhi (1645 A.D.). The front 3 halls were renamed the Hall of Supreme Harmony, the Hall of Central Harmony and the Hall of Preserved Harmony, but few of the internal palaces were renamed. The construction was completed the next year. The internal palaces were repaired in the 12th year of the reign of Emperor Shunzhi (1655A.D.). 3 major palaces were repaired in the 30th year of the reign of Emperor Qianlong (1765 A.D.). In 1774, Emperor Qianlong ordered to build Wenyuan Pavilion behind Wenhua Hall to conserve *Siku Quanshu* (Complete Library of the Four Branches) edited by imperial order. The Hall of Heavenly Purity and the Hall of Union and Peace suffered a fire and were repaired in the 2nd year of the reign of Emperor Jiaqing (1797 A.D.), and were completed the next year. Since then the palaces haven't experienced any large repair.

There was a special architecture in the palace. The Imperial Garden at the rear of the central axis, was built initially during the reign of Emperor Yongle, and enlarged during the reigns of Emperor Jingtai and Emperor Wanli. The eye-appealing rockerys, verdant plants, ancient cypress and vines, were all inherited from the Ming Dynasty.

After Sun Yat-Sen led the Chinese Revolution of 1911 and overthrew the Qing Dynasty, Qing's royal family was permitted to live in the inner court according to an agreement. In 1914, the front 3 halls began serving as an antique exhibition hall, representing the start for the Imperial Palace to be a museum. Feng Yuxiang, a warlord, expelled the abdicated emperor of the Qing Dynasty out of the imperial palaces in 1924. During the Anti-Japanese War, lots of rare cultural relics were migrated southwards for protection. Since the founding of the People's Republic of China, the Imperial Palace has been a museum.

## 2. Architecture and Art of Palaces in Beijing

### (1) The past and today

The palaces in Beijing occupied a smaller acreage than those in early period and middle period of ancient China. However, the palaces in early period relied more on the beauty of natural environment. The palaces in Beijing in the middle and late periods basically concentrated within the royal city. However, the palace layout in such a period was more precise, concentrated and reasonable. It made full use of the space even within a limited scope and used architectural techniques to foil the environment. Various architectural artistic conceptions were created through cadence of architectural techniques in a limited space, in order to fulfill the requirements for the functions of the architecture and rely mainly on the artistic appeal of the architecture itself. The Forbidden City was the essential works among Chinese architecture during thousands of years.

### (2) China and foreign countries

Ancient Chinese architecture had their own style. Marco Polo admired the palaces in Dadu very much in his *The travels of Marco Polo*, which surprised the western world. Besides, the Forbidden City in Beijing, as a typical of Ancient Chinese architecture, is the world's existing largest and most complete palace complex with the longest history. Their high achievement in architectural technology and art directly impact palace architectures in Japan, Korea and other countries.

### (3) The Forbidden City, a bright pearl

The existing palaces in the Forbidden City of the Ming and Qing Dynasties, like their counterparts in other past dynasties, followed the rites and city systems fixed by Kings of the Zhou Dynasty in planning and layout. They further standardized the modulus design system for palace construction since the Tang and Song Dynasties, adopted Geomantic Omen theory in traditional Chinese culture, and came down in one continuous line with imperial palaces of past dynasties. As for the architectural artistic conception, they represented the culmination of using architecture complex to foil stateliness and holiness of the emperor. As the highest achievement of large architecture complex in the late of feudal society in China, they were more mature in architectural technology and featured highly standardized design and construction.

First, the Forbidden City of the Ming and Qing Dynasties brought traditional Chinese axis symmetry to the perfection. The main palaces in the Forbidden City were on a 1.6 kilometers long central axis from the south to the north. They consisted of sequential and symmetrical enclosed spaces and a series of corridors and palaces. Such an unfolded architectural sequence foiled the grandiosity, stateliness and holiness of 3 major halls. Besides, the central axis of the Forbidden City was the same as that of the whole Beijing city, from Yongdingmen to Daqingmen, northwards through palace axis to Drum Tower and Bell Tower. The axis went through the whole city, extending the axis of palace city southwards and northwards and increasing the palaces' grandiosity of axis symmetry. It was the world's longest city axis.

Second, these palaces used colors successfully. The Forbidden City of the Ming and Qing Dynasties had yellow glazed tiles and red walls. It was located at the central of the city. Seen from far away, a high and gorgeous architecture complex distinguished itself, which was a distinct contrast against low and grey architecture cross the whole city. The contrast of color realized the magnificent and gallant effect required for the palaces.

Third, these palaces created architecture artistic imageries successfully. The Forbidden City created various atmospheres successfully with architecture, having a stately, lofty, grand, and luxury effect of the imperial palaces and inspiring various psychological changes in visitors, in order to embody sovereign imperial power and holiness and sense of mystery of the emperor.

# 故宫
# The Palace Museum

　　故宫是我国明清两代的皇宫，又称紫禁城，先后有明、清两代的24位皇帝在此登基执政，同时故宫也是世界上现存规模最大、最完整的古代宫殿建筑群。1961年故宫被公布为全国重点文物保护单位，1987年被列入世界文化遗产名录。

　　紫禁城始建于明永乐四年(1406年)，历时14年完工，之后又经明清两代多次重修和扩建，但仍保持着原来所具有的布局和建筑风格。

　　紫禁城的由来是取天上的紫微星垣的"紫"字，因为紫微星垣居于天市中心，又因为皇帝的皇宫是"禁地"，故名紫禁城。紫禁城的宫殿沿着一条南北向中轴线排列，并向两旁展开，左右对称。这条中轴线不仅贯穿紫禁城内，而且向南北延伸，南达永定门，北至鼓楼、钟楼，贯穿了整个城市，总长度7.8公里。

　　紫禁城南北长961米，东西宽753米，占地72万平方米，总建筑面积17万平方米，共有宫殿8700多间，均为木结构、黄琉璃瓦顶、青白石台基，并饰以金碧辉煌的彩画。紫禁城四周是高大宏伟的红色宫墙，长达3400米，宫墙四面正中各辟一门，南门名午门，东门名东华门，相对之西门称西华门，北门名神武门；宫墙的4个角各矗立一座风格独特、造型秀丽的角楼；宫墙外环绕着宽52米的护城河，与高大的宫墙、角楼构成了一个宏伟、森严的城堡。

　　紫禁城的宫殿分为外朝区和内廷区两个大的部分，建筑风格迥然不同，营造出的境界也各不相同。

　　外朝区是皇帝举行登基等大典、行使国家大权的场所，建筑高大、威严、宏敞。午门是紫禁城的正门、外朝的开端。过午门后眼前豁然开朗，前面是5座雕琢精美，形似玉带的汉白玉石桥，称为内金水桥，内金水桥后面是太和门，明清两代多位皇帝在此处理朝政，故有"御门听政"之说。过太和门即是以太和、中和、保和三大殿为中心，文华、武英两殿为两翼的外朝区主体建筑群。

　　保和殿之后为内廷区，是皇帝日常处理政务、居住、和后妃游玩、奉神之处，建筑相对于前朝区稍小且紧凑。内廷区的主体建筑是皇帝常朝的乾清宫、皇帝的正寝宫交泰殿和皇后的正寝宫坤宁宫，统称"后三宫"。

从天安门俯瞰紫禁城
The Bird View of the Forbidden City from the Gate of Heavenly Peace

　　乾清宫东西各有六组院落，自成体系，即东六宫和西六宫。这些宫院的每个院落均由前后殿、东西庑的标准格局组成，其中储秀宫、翊坤宫、体元殿构成的院落是典型的后宫寝室形式，居住后妃，室内布置陈设极尽豪华。东六宫以南有奉先殿、斋宫、毓庆宫，西六宫以南有养心殿。内廷另有三座花园，即紫禁城中轴线结尾处的御花

园、宁寿宫和养心殿西侧的宁寿宫花园、慈宁宫前的慈宁宫花园。

外朝、内廷区以外还有外东路和外西路两路建筑。外东路南部为皇子居住的南三所，北部为乾隆皇帝养老的宁寿宫。外西路有皇太后居住的慈宁宫、寿康宫和英华殿等佛堂建筑。

故宫规制宏伟，布局严整，建筑精美，富丽华贵，并收藏有明清两代皇宫流传至今的许多稀世珍宝，是我国古代建筑、文化、艺术的精华，承载着中华民族千百年来最深沉的积淀。

The Palace Museum, known as the Forbidden City, was the imperial palace during the Ming and Qing Dynasties. Altogether twenty-four emperors reigned and ruled here. It is the largest and most complete group of ancient buildings standing in the world. Listed by the State Council as a national key relic under special preservation in 1961, the Palace Museum was also inscribed in the World Heritage List of the United Nations Educational, Scientific and Cultural Organization (UNESCO) in 1987.

Construction of the Forbidden City began in the 4th year of the reign of Emperor Yongle of the Ming Dynasty (1406 A.D.). It was completed 14 years later in 1420. Many of the buildings had been repaired and rebuilt during the Ming and Qing Dynasties, but their basic form and layout remain in their original state.

It is a location endowed with cosmic significance by ancient China's astronomers. Correlating the emperor's abode, which they considered the pivot of the terrestrial world, with Ziweiyuan (the Pole Star), which they believed to be at the center of the heavens, they called the palace the Purple Forbidden City. Once inside, visitors will see a succession of halls and palaces spreading out on either side of an invisible central axis line 7.8-kilometer-long which begins in the south at Yongdingmen Gate (no longer extant) in the former outer city wall; further north, it passes through the Forbidden City. It continues north and ends at the Drum and Bell Towers. It is a fundamental feature of Beijing's layout and splits the city into approximate halves.

Measuring 961 meters from north to south and 753 meters from east to west, the Forbidden City covers an area of 720,000 square meters, having 8,700 buildings. It is a magnificent sight, the buildings' glowing yellow roofs against vermilion walls, not to mention their painted ridges and carved beams, all contributing to the sumptuous effect. The rectangular city is encircled in a 52-meter-wide moat and 3,400-meter-long red city wall which has one gate on each side: the Meridian Gate to the south, the Gate of Martial Spirit to the north, the East Glorious Gate to the east and the West Glorious Gate to the west. There are four unique and delicate structured corner towers

overlooking the city inside and outside on the four corners.

The Imperial Palace is divided into two ceremonial areas: the Outer Palace and the Inner Court.

The Outer Court was the place where emperors reigned or executed their supreme powers over the nation. The Meridian Gate is the southern entrance of the Forbidden City. Through the Meridian Gate, one can see five bridges called the Inner Golden Water Bridges. The bridges are well decorated with marble balustrades carved with motifs of dragons and phoenixes. Further north in the center, it is the Gate of Supreme Harmony. In the Ming and Qing Dynasties, many emperors used the gate as a place to listen to reports and issue proclamations. This was known as "holding court at the Imperial Gate". The three main halls of the Outer Court, the Hall of Supreme

Harmony, the Hall of Central Harmony and the Hall of Preserved Harmony sit in line inside the gate.

Behind the Hall of Preserved Harmony is the Inner Court that was the residential area of emperors and the imperial household, as well as where emperors lived or dealt with routine state affairs. The principal structures are Qianqinggong (the Palace of Heavenly Purity), Jiaotaidian (the Hall of Union) and Kunninggong (the Palace of Earthly Tranquility) straddling the central axis, surrounded by the Six Palaces of the East and West and the Imperial Garden to the north. Other major buildings include Fengxiandian (the Hall for Worshipping Ancestors) and Zhaigong (the Palace of Abstinence) on the east, and Yangxindian (the Hall of Mental Cultivation), Yongshougong (the Palace of Eternal Longevity) and

Yikungong (the Palace of the Queen Consort) on the west. These contain not only the residences of emperors and his empresses, consorts and concubines but also the venues for religious rites and administrative activities.

With its carefully-designed layout, splendid structure and decoration, clear and rigid hierarchy, flexible spatial composition, the Place Museum embodies the extensiveness and profundity of traditional Chinese art, and represents the highest achievements ancient China was able to make in architectural engineering and art.

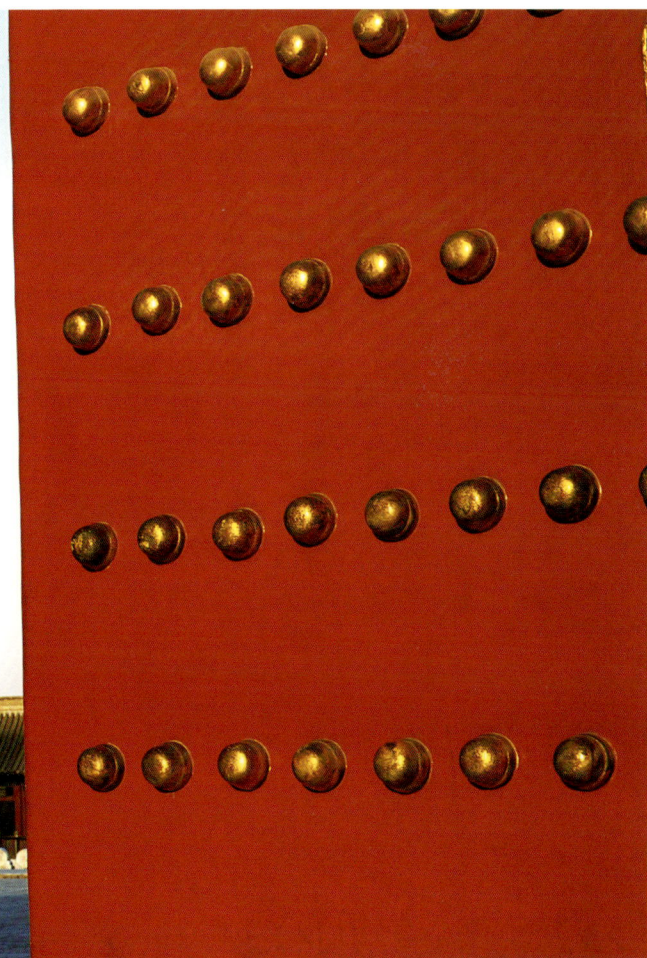

阙右门与午门
The Queyou Gate and the Meridian Gate

午门
The Meridian Gate

午门平面呈倒"凹"字形，正面城台上建三座重檐庑殿式门楼，左右延伸的城台上各建一座重檐四角攒尖顶方亭，形成五座城楼，俗称五凤楼。明代时皇帝每逢正月十五，均在午门张灯结彩赐宴百官，并且在立春日赐春饼，端午日赐凉糕，重阳日赐花糕。此外逢国家征战，皇帝在此接受凯旋的军队献上的战俘。另外，午门外还经常责罚大臣，称为"廷杖"，有的甚至被打死，所以民间流传有推出午门斩首的传说。一到午门前就让人不自觉地产生一种非常森严的感觉，人站在下面会感觉自己非常渺小。

午门背面
The North of the Meridian Gate

太和门广场
The Square in Front of the Gate of
Supreme Harmony

太和门
The Gate of Supreme Harmony

太和门为外朝三大殿的正门，明初称奉天门，也是紫禁城最宏伟、高
大的宫门。

太和门彩绘
Coloured Paintings of the Gate
of Supreme Harmony

太和殿广场
The Square of the Hall of Supreme Harmony

太和殿，又称金銮殿，是举行皇帝即位、诞辰以及节日庆典和出兵征
伐等重大国典的地方。初建于明永乐十八年(1420年)，康熙三十四年
(1695年)重新修建。一过太和门正北是太和殿广场，广场可容纳万
人，大殿位于广场北侧，建在三层汉白玉台基之上，台基四周环绕着
云凤望柱，雕刻吐水螭首，站在太和门远望太和殿如天上宫阙一般。
登殿的石阶正中是巨型石雕蟠龙云海御路石。台上摆放有铜龟、鹤，
象征皇家江山永固和万年长青；摆放的测量时间的日晷象征皇帝勤勉
为国；还有计量器皿嘉量，象征皇帝公允持国。台上还有18尊只用于
庆典时焚烧香料的大铜鼎。

太和殿线图
The Hall of Supreme Harmony

太和殿面阔十一间，进深五间，重檐庑殿顶，高35米，宽63米，面积
2377平方米，是我国现存等级最高、体量最大的古代木构建筑。

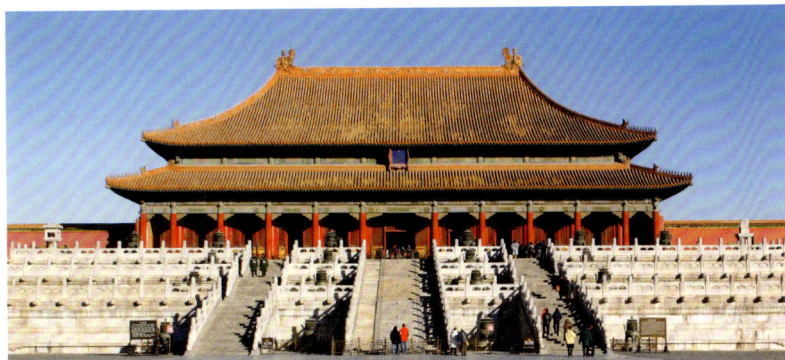

太和殿
The Hall of Supreme Harmony

太和殿远望
The Hall of Supreme Harmony
Viewed from Far

太和殿吻兽
Wen at the End of the Main Ridge of the Hall of Supreme Harmony

太和殿脊兽
Decorative Figurines of Immortals and Mythical Beasts on the Eave
Extensions of Roof Corner of the Hall of Supreme Harmony

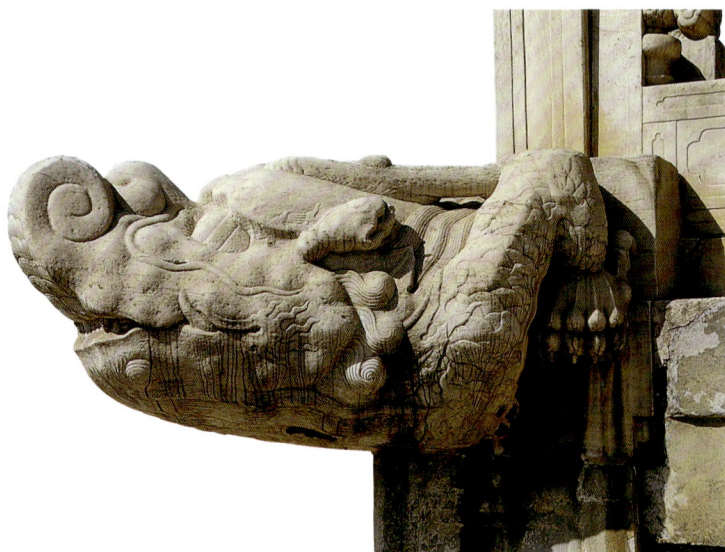

螭首
The Head of Chi, a Kind of Hornless Dragon

三台
The Three-Tiered Terraces

三大殿建在一座高8.13米的巨大"土"字形台基之上，为三层汉白玉石台基，台基四周雕有上千个螭首(螭是龙的一种)。这些螭首的一个功能是美观，另外它与台基上的排水管道相通，还有排泄台基雨水的功能。雨水从上千个螭首嘴中喷出，所以每逢大雨时，三大殿便形成了"千龙吐水"的一大奇观。

太和殿嘉量
Jialiang on the Platform of the
Hall of Supreme Harmony

太和殿铜鹤
Bronze Crane on the Platform of
the Hall of Supreme Harmony

太和殿铜龟
Bronze Tortoise on the Platform of
the Hall of Supreme Harmony

太和殿日晷
Stone Sundial on the Platform of the Hall of Supreme Harmony

太和殿廊柱
Columns at the Front Veranda of
the Hall of Supreme Harmony

太和殿内景
Interior of the Hall of Supreme Harmony

太和殿内共有72根大柱，其中顶梁大柱最粗最高，直径为1.06米，高为12.7米。明代用的是楠木，采自川、粤、云、贵等地；清代重建后，用的是松木，采自东北三省的深山之中。太和殿的中央设楠木镂空透雕龙纹的金漆基台，上设九龙金漆宝座，宝座背后有雕龙金漆屏风，宝座两侧又有6根贴金盘龙龙井柱。殿顶全部为金龙图案的井口天花，宝座正中为蟠龙衔珠藻井，梁枋上也遍饰金龙和玺彩画，使整个大殿形成万龙竞舞的气氛。太和殿内地面共铺两尺见方的大金砖4718块，金砖并不是真正用黄金制成，而是采用苏州一带特有的土，经过3年时间，数十道工艺加工后，烧制成的特殊的砖，它质地坚硬，密度很大。

太和殿宝座
The Throne in the Hall of Supreme Harmony

太和殿轩辕镜
The Mirror of the Yellow Emperor in the
Hall of Supreme Harmony

三大殿侧面
The Side of Three Great Halls

中和殿
The Hall of Central Harmony

中和殿平面正方形，深广各五间。屋顶为四角攒尖顶、顶上的铜胎鎏金
宝顶犹如一颗巨大的宝珠，在阳光下熠熠闪光。中和殿面积仅有580多平
方米，小巧端庄，点缀在庄严宏伟的太和殿与保和殿之间，平添了一抹
轻灵秀丽之感。 明清两代，中和殿是皇帝在前往太和殿途中的小憩之
处，并先在此接受内阁、礼部及侍卫执事人员的朝拜。

中和殿正立面图
The Elevation of the Hall of Central Harmony

中和殿内景
Interior of the Hall of Central Harmony

堂皇的宫殿门窗
The Magnificent Doors and
Windows of the Palace

保和殿正立面图
The Elevation of the Hall of Preserved Harmony

保和殿
The Hall of Preserved Harmony

保和殿面阔九间，进深五间，重檐歇山顶。平面中减去了殿内前面的6根大柱，使殿内的视线豁然开朗。明代，皇帝举行典礼前在此更换礼服。清代每逢佳节，皇帝在此宴请蒙古、西藏等外藩王公贵族和京中文武大臣，清代中后期也是举行殿试的场所，进京赶考的各地举人中了进士之后，由皇帝在这里最后出题考察、选定殿试三甲，即状元、榜眼和探花。

保和殿内景
Interior of the Hall of Preserved
Harmony

从三大殿看后宫
Looking at the Inner Court from
Three Great Halls

保和殿后大丹陛
Imperial Way Stair Slab Carved with Clouds and Dragons on the
Ramps in Front of the Hall of Preserved Harmony

保和殿后阶陛中间有一块雕刻着云、龙、海水江崖的御路石，人们称之为云龙石雕。这是紫禁城中最大的一块石雕，长16.57米、宽3.07米，厚1.7米，重为250吨。原为明代雕刻，清代乾隆时期又重新雕刻。图案是在山崖、海水和流云之中，有9条口戏宝珠的游龙，它们的形象动态十足，生机盎然，是一幅巨大的石雕精品。

乾清门广场
The Square in Front of the Gate of
Heavenly Purity

乾清门铜狮
The Bronze Lion in Front of the
Gate of Heavenly Purity

乾清宫
The Palace of Heavenly Purity

乾清宫是一座面阔九间、进深五间的重檐庑殿顶建筑，殿内外梁枋饰以金龙和玺彩画，富丽典雅。乾清宫先后有明代的14位皇帝和清代的顺治、康熙两位皇帝以此为寝宫。乾清宫正殿悬挂着由清代顺治皇帝御笔亲书的"正大光明"匾一块，自雍正皇帝执政开始，这个匾的背后便藏下了天下关注的秘密——"建储匣"。雍正朝鉴于皇子之间相互倾轧争夺皇位的情况，开始采取秘密立储君的办法，即皇帝生前不公开立皇太子，而秘密写出皇位继承人的文书，一式两份，一份放在皇帝身边；一份封在"建储匣"内，放到"正大光明"匾的背后。待到老皇帝死后，将匾额后面的一份取下和皇帝秘藏在身边的一份对照验看，经核实后册立新君。乾隆、嘉庆、道光、咸丰四帝，都是按此制度登上宝座的。清代后期，由于咸丰皇帝只有一个儿子即同治皇帝，而同治和光绪皇帝都无后，这种秘密立储的办法才失去其意义。

江山社稷金殿
The Square Pavilion of the Gods
of Land and Grain

乾清宫内景
Interior of the Palace of
Heavenly Purity

正大光明

表正萬邦慎厥身脩思永

克寬克仁皇建其有極

交泰殿内景
Interior of the Hall of Union

交泰殿为单檐四角攒尖顶方殿，深广各三间，铜镀金宝顶，似中和殿
而缩小。开始为皇帝大婚的洞房，清代乾隆朝开始将象征皇帝权力的
25方玉玺存放在此处。

坤宁宫东暖阁
The Eastern Chamber of Warmth in the Palace of Earthly Tranquility

坤宁宫面阔九间，进深五间，高约22米，重檐庑殿顶，梁枋饰和玺彩
画。在明代时为皇后的正寝宫，清代时则改成皇帝的洞房和供奉满族
信奉的萨满教神灵之处。

养心门
The Gate of Mental Cultivation

养心殿
The Hall of Mental Cultivation

养心殿是皇帝居住和处理日常政务的地方，其正间为皇帝接见官员、商议朝政的地方，西间是皇帝阅览奏折和议事处，东间在同治、光绪两帝执政期间，是慈禧太后垂帘听政的地方。

储秀宫
The Palace of Gathering Elegance

慈禧太后曾居住在储秀宫内，她50岁寿辰时耗费白银125万两重修，
是紫禁城内最豪华的宫室之一。

储秀宫门饰
The Decoration on Doors of the
Palace of Gathering Elegance

储秀宫内景
Interior of the Palace of Gathering
Elegance

順時施宜

乙酉中秋

乾隆花园楔赏亭
The Pavilion of the Ceremony
of Purification

乾隆花园耸秀亭
The Songxiu Pavilion

乾隆花园古华轩
The Guhua Hall

堆秀山
The Hill of Accumulated Beauty

千秋亭
The Pavilion of Thousand Autumns

断虹桥
The Duanhong Bridge

元代石雕戗兽
Stone Beast of the Yuan Dynasty

角楼
The Corner Tower

角楼内部结构线图
Cross Section of the Corner Tower

神武门
The Gate of Martial Spirit

角楼
The Corner Tower

坛庙

# ALTARS
# AND TEMPLES

# 北京的坛庙

坛庙建筑是人类自然崇拜和祖先崇拜的一种体现。从浙江余杭反山和瑶山、内蒙古包头阿善、辽宁喀左东山嘴、辽宁建平牛河梁等地发现的新石器时代的祭坛，到周代的祭天建筑——明堂和圜丘；从汉代成书的《周礼·考工记》的皇宫要建"左祖右社"的记载，到至今仍多数留存的明清北京城内著名的"九坛八庙"，坛庙建筑一脉相承，源远流长，长盛不衰，形成了中国古代建筑中很有特色的一大类——坛庙建筑。北京作为五朝古都，中国封建社会后期的政治、文化中心，坛庙建筑无论是规模，还是数量在世界上都是独一无二的。那一座座曾经神秘和肃穆的坛庙，为古都北京增添了无数的神韵。

## 一、北京坛庙建筑的发展概况

### 1.商周至金代北京的坛庙建造情况

周初至战国末期，北京作为诸侯国燕和蓟的都城，燕、蓟所建"沮泽"，是为坛庙雏形。辽代，北京作为陪都南京，辽太祖时就在北京建立了家神庙，而且以后逐渐形成制度化。

辽亡金兴，金代统治者相对于辽代统治者更多吸收了汉文化，而且把宋汴梁的许多礼器都运回使用。金人本来有拜天的习俗，金太宗继位后，仿照汉人礼俗设位而祭，金天德以后，开始有郊祀，至金大定、明昌年间郊祀制度齐备。城外设天、地、日、月四坛，分别位于南、北、东、西四方。此外，金还建立了社稷坛、祭拜祖先的太庙及安置皇帝画像和生前所使用之物的原庙。

### 2.元代北京的坛庙

元朝兴起于朔漠，也有拜天的传统，但是按照蒙古族的方式，仪式简单而淳朴，也没有正式的祭坛，即使至元十二年(1275年)忽必烈接受"尊号"也只是在大都城南门丽正门外东南七里设台祭天地神位，但这也不是传统意义上的"南郊礼"，只是临时性地对天地的预告，祭台也不是按照圜丘形式建造，祭祀仪式也是按照蒙古族的旧俗。直至

成宗大德九年(1305年)，才因多天灾而在大都城南郊正式建坛合祭皇天上帝、皇地祇、五方帝。至元三十年(1293年)，在和义门内稍南建立了社稷坛，占地40亩，社、稷二神分坛而祭，二坛并排而列，同在一道墙墙内，社坛在东，稷坛在西。至于地坛、日坛、月坛等，元代则根本没有建立。至元四年(1267年)诏建大都太庙，至元十七年(1280年)建成。

太庙位于宫城东北齐化门内，庙制为前殿后寝，大殿七间，内分七室，仿照金代宗庙形制。元代在大都还修建了孔庙，明清一直延续使用。

### 3.明代北京坛庙格局的形成

明朝建立后，朱元璋非常重视礼乐制度，制定了包括大祀、中祀、小祀等典礼制度，在南京钟山之阳建圜丘以祭天，钟山之阴建方泽以祭地。洪武十年(1377年)改天地分祀为合祭，建大祀殿。明成祖朱棣夺得皇位后，计划迁都北京。坛庙建设是新都的重要组成部分。永乐十五年(1417

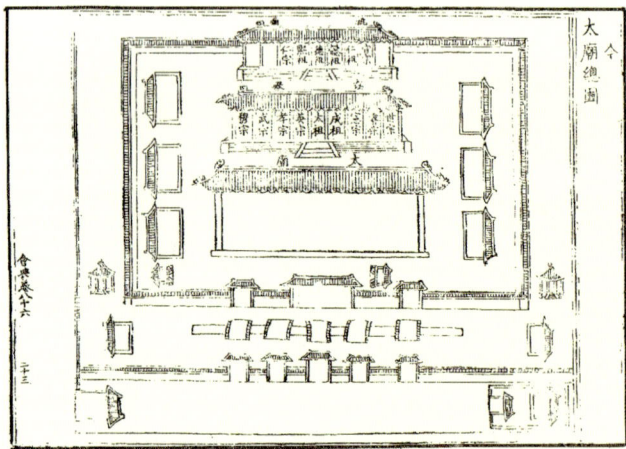

太庙总图[《大明会典》(今)卷八十六]

年)六月郊坛建设动工，永乐十八年(1420年)北京天地坛完工。同年，明成祖将都城由南京迁到北京。翌年成祖以北京郊社宗庙及宫殿建成，亲自到太庙祭祀列祖列宗。10天后又大祀天地于南郊。

明成祖仿照南京建成的北京坛庙格局，即太庙与社稷坛遵从《周礼·考工记》"左祖右社"原则，位于紫禁城南左右，天地坛和山川坛作为合祭天地众自然神的郊坛分列正阳门外左右，这种格局沿用了160余年，至明世宗嘉靖年间，这种祭祀局面才被打破。

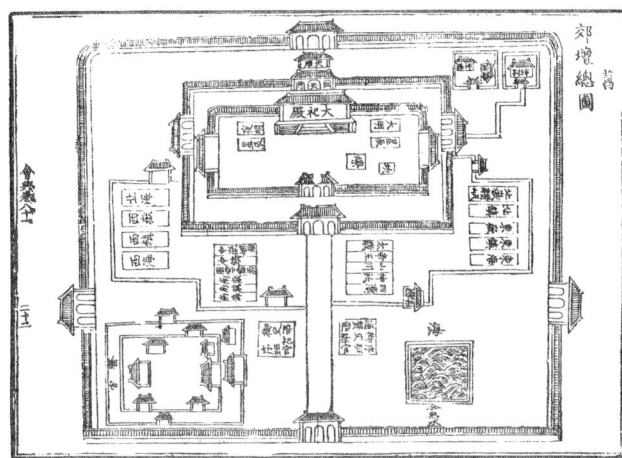

郊坛总图 [《大明会典》(旧) 卷八十一]

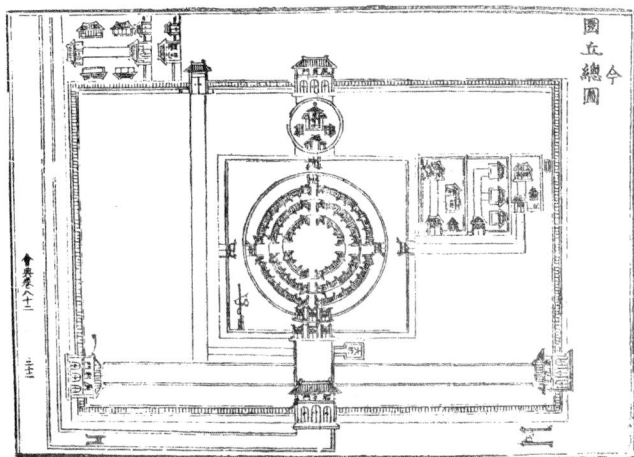

圜丘总图 [《大明会典》(今) 卷八十二]

正德十六年(1521年)三月，明武宗崩，无嗣，由兴献王的儿子朱厚熜承继大统，年号嘉靖。朱厚熜由藩王世子入继皇位，为了给自己正名，欲将其亲生父亲尊为皇考，遭到群臣的激烈反对，引起了明代有名的"大礼仪"之争，世

宗甚至在众大臣面前痛哭流涕地表示不愿意再做皇帝，要护送着自己父亲的灵位回家，众大臣无奈只好暂时让步。世宗在全面掌握了政权以后以高压手段平息争议，而与"大礼仪"之争相伴而生的是嘉靖帝对祀典的全面更定。嘉靖九年(1530年)，朱厚熜决定恢复明太祖初制，即天地分祭的制度。于城南、北、东、西分别建造圜丘(天坛)、方泽坛(地坛)、日坛、月坛分祀天、地、日、月。建成后的郊坛，天坛在大祀殿之南建新建了圜丘，圜丘北即过去的天坛所在地大祀殿。嘉靖十一年(1532年)，在圜丘坛外泰元门东建起崇雩坛，用以孟夏时节皇帝举行祈雨大典之用，但自建成后即荒疏而未常用，而在圜丘上举行，到了清代乾隆朝将之拆除。方泽坛坛制与现在的地坛基本相同。朝日坛在朝阳门外；夕月坛在阜成门外；此外还在山川坛内建造了天神地祇坛，奠定了今天先农坛的格局。

四郊分祀以后，大祀殿废而不用，嘉靖十七年(1538年)六月，前扬州府同知丰坊上书皇帝，请求皇帝为其父献皇帝建明堂配天而祭，即举行季秋大享礼。这正符合嘉靖帝的心意，命令拆除大祀殿。嘉靖十九年(1540年)在大祀殿旧址上建三重檐攒尖顶的大享殿，即今祈年殿的前身。嘉靖帝死后，大享礼即废，大享殿徒有其名。

明嘉靖三十二年(1553年)，北京城南的外城建成，原南郊天坛实已不属郊外而被围在北京城内。天坛位于外城永定门至内城正阳门即北京城中轴线南段以东，以西为先农坛。为壮观瞻，使两坛坛墙更加完善，同时也与整个北京城规划布局相合谐，天坛增筑外坛墙。形成天坛、先农坛两坛夹天街对峙的布局，蔚为壮观。天坛外坛墙建成后，仅西向设门，坛域扩大为273万平方米。但连接南北两坛的轴线不居中，而位于中线偏东位置，打破了传统的主要建筑位于中轴线上的做法。这与整个北京城西城略大、东城略小的特点相一致。轴线偏东丝毫未使天坛建筑布局受损，反而使得由西向东进入天坛的人更觉得坛域深远广阔，崇高意境更为突出。

明嘉靖朝对坛庙的一系列改革，使得北京的坛庙格局已经基本形成。

### 4.清代对坛庙建筑的变革及最终定型

清入关前，在盛京(即今辽宁沈阳)就建有圜丘、方泽坛，祭告天地。顺治元年(1644年)，清军入关后便立即遣使臣祭告北京的天坛。清初的顺治、康熙、雍正三朝基本上沿袭了明代的坛庙，只是在礼仪上更加严格、隆重。仅康熙朝(1662－1722年)就举行了祭天大典50次，祈谷大典49次，其他诸如京师地震、为祖母祈寿、平叛告捷等告祭皇天上帝礼仪9次。

乾隆皇帝继位后，由于经过清前期的积累，国库丰盈、政治安定，有能力再次大规模修建坛庙建筑。

乾隆八年(1743年)，因天坛斋宫破旧决定整修，乾隆十二年(1747年)，因天坛内外坛墙年久损坏严重，为整齐划一，将原土墙拆修。同年奏准拆除崇雩坛，祈雨礼改在圜丘进行，拆下的城砖在修理天坛内外垣时使用，从而拉开了清高宗大兴土木，改、扩建坛庙建筑的序幕。

乾隆十四年(1749年)，因圜丘坛上张设幄次陈祭品处过窄，于是决定扩建圜丘。乾隆十五年(1750年)，又将砖色不一的大享殿三层坛面俱用金砖墁砌。同时乾隆认为大享殿前后两庑是前明合祭所建，而今合祭之礼已不举行，并且两庑前重九间，后重七间又参差不齐，于是下令将后一重拆去。乾隆十六年(1751年)七月初一日，改祈谷坛旧有题额"大享殿"为"祈年殿"，由明合祀天神地祇向清单纯祈谷于上帝的变化。祈年殿两庑绿瓦改为青色，而祈年殿门三座及随门围垣，因为离坛稍远，仍照旧制覆盖绿瓦。乾隆十七年(1752年)，改建皇穹宇，将皇穹宇重檐式殿顶改作单檐式，地面用青石铺墁，围墙墙身及槛墙用临清(今山东临清)城砖砌成。此城砖以"敲之有声，断之无孔"著称于世，皇穹宇围墙就是举世闻名的"回音壁"。乾隆十九年(1754年)，在天坛西门外垣之南建门一座，称"圜丘坛门"，原来的西门称之为"祈谷坛门"，形成了天坛南北两坛，规制严谨的格局。至此，天坛最终形成了今天所见到的面貌。

其他祭坛在乾隆时期亦得到修缮，或拓展坛制，或更新瓦色，或修筑墙垣，并补植树木，使祭坛殿台鼎新，愈加肃穆庄严。先蚕坛也在乾隆九年(1744年)建成，坛址位于西苑东北隅(今北海内)，临太液池，有先蚕坛、观桑台、桑园、亲蚕殿、神库、神厨、宰牲亭、蚕署、蚕所等建筑。

乾隆一朝，北京祭坛发生了巨大变化，色彩更鲜明浓烈，建筑更圣洁崇高，大量树木的补植，使祭坛氛围更肃穆清朗。

清代后期由于国衰君庸、外忧内患，坛庙基本上也没有再进行大规模修缮，祭祀活动也不再像中前期那么严格了，尤其是清代晚期皇帝往往不再亲自前往祭坛斋戒、祭拜，而是派王公代为前往，祭祀活动愈加荒疏。光绪十五年(1889年)，天坛祈年殿遭雷击烧毁，次年下令重建，至光绪二十二年(1896年)竣工。

1900年，八国联军攻占北京后，慈禧太后携光绪皇帝西逃，昔日皇家神坛遭到大量破坏，祭坛内陈设、祭器、礼器及库存物品被洗劫一空。光绪三十四年(1908年)，除太庙以外停止所有坛庙祭祀，祭期来临，或遣官恭代行礼，或竟至停祀。1911年，辛亥革命推翻清王朝，祭坛制度随之废止。

## 二、北京现存坛庙的情况

北京现存的坛庙从皇家的坛庙到名人祠庙，基本上保存了明清以来的格局，可以分为以下几类：

### 1.天神坛

主要包括天、日、月、风、云、雨、雷等，这些坛庙基本上保存完整，而且都是皇家坛庙，占地广阔，建筑等级也很高。

### 2.地神坛

主要包括地、先农、先蚕、岳、镇、海、渎、名山大川等神的坛庙。

### 3.祠庙

北京的祠庙是数量最多的，从皇家供奉祖先牌位的太庙，供奉历代帝王先圣的历代帝王庙，祭祀先师孔子的孔庙，祭祀历朝名臣贤相的祠庙，如文天祥祠、于谦祠、袁崇焕祠等，到一个家族的祠堂、家庙，可谓数量宏大。家族祠堂或家庙一般占地相较于皇家祠庙和功臣祠庙规模要小得多，多为一两进院落，甚至只有一座房子。

## 三、北京坛庙的建筑、艺术成就

宏大的坛庙建筑和盛大的祭祀典礼活动，使得北京的礼制建筑规模空前，成为世界上唯一一个拥有如此大规模坛庙建筑的东方古都。

### 1.建筑技术上

北京现存的古代坛庙是中国古代社会最后的巅峰之作，各种传统技术完全成熟。如天坛皇穹宇的回音壁，成功地利用声学原理；又如站在圜丘坛的天心石上说话，感觉声音异常洪亮，也是声学原理的成功运用。

### 2.建筑艺术上

①成功地将中国古代的宇宙观、古代哲学与建筑相结合，使建筑本身处处体现了中国古代的哲学观和宇宙观。如祈年殿最里圈的4根龙井柱寓一年的春夏秋冬四季，内檐金柱一圈12根寓一年12个月，外檐柱一圈12根寓意一天12时辰，内檐和外檐柱相加寓一年二十四节气，内外总计28根寓意周天二十八星宿；又如圜丘坛围绕天心石四周环砌扇形台面石，第一圈9块，依次往外每圈递增9块，直至"九九"81块，寓意中国古代认为的"九重天"。而这种建筑的象征的手法在北京所有的坛庙建筑中都有体现，似乎那古老神坛的一砖一瓦都蕴含着中国传统文化的，漫步其中，仿佛徜徉于中国古代文化的宝库。

②建筑意境的运用达到了中国传统建筑的高峰。建筑意境上的抑、扬、顿、挫、借等取景方法和建筑的高低搭配、色彩对比等各种手法运用得淋漓尽致，为各座古老的神坛、圣庙创造出了功能上要求的宁谧、浩远、肃穆的意境。如各个坛庙四周种植茂密的柏树，从坛内高处向四周望去满眼葱绿，顿觉似在天宇，望来天高地迥、宇宙无穷，从而油然而生对天地日月的崇敬之情。又如太庙的建筑为了创造出肃穆的意境，殿宇安排得异常高大紧凑，再加上宏伟的基座，衬以周围矮小、低平的建筑，置身其中顿觉压抑、肃穆，崇敬之情油然而生。

③建筑语言的运用。北京坛庙所采用的建筑语言主要有：天为阳，地为阴的阴阳学说，所以天坛选址在南面，就阳位，地坛在北，就阴位。在建筑细部处理上，系统地运用数字这一带有隐喻性或象征性的建筑语言，天坛、日坛建筑用一、三、五、七、九阳数，地坛、月坛建筑用二、四、六、八阴数。中国古代"天圆地方"说，认为"天圆如张盖，地方如棋局"。所以天坛的圜丘和祈年殿等祭祀建筑都为圆形，而地坛的方泽坛等祭祀建筑都为方形；天坛的坛墙为南方北圆；天坛的圜丘四周墙墙内层为圆形外层为方形；方泽坛的方、日坛拜台的圆形墙墙、月坛方形墙墙诸如此类也都寓意天圆地方。古人的"坛而不屋"说，认为天地日月的本性纯朴自然，无需华饰，所以应依照天地本性用自然的东西表达对其的敬意，于是积土为坛，不加雕饰。后来由于逢风雨不方便祭祀，明代时干脆建了大祀殿祭祀，所以嘉靖帝认为这不符合古制，重设了圜丘。此外，各个坛庙的坛台都没有雕饰。

置身在北京这些古老的神坛，感受它厚重的历史感和绝伦的建筑将带给我们无尽的精神享受。

# Altars and Temples in Beijing

The very characteristic buildings among the Chinese ancient complex are architecture of altars and temples. As the ancient capital for five dynasties and the politic and cultural center in later period of feudal society, Beijing has the unique altar and temple architecture, in terms of whether scale or number.

## 1. The Outline on the Development of Altar and Temple Architecture in Beijing

### (1) The altar and temple construction in Beijing during the Shang, Zhou and Jin Dynasties

From the early Zhou Dynasty to the late Warring States Period, Juze, as an incipient form of temples, was established in Beijing, the capital of the Yan State and Ji State. In the Liao Dynasty, Beijing became an auxiliary capital called Nanjing, where an ancestral temple was established during the reign of Emperor Taizu.

During the Dading and Mingchangs' reigns of the Jin Dynasty, four altars of heaven, earth, sun, and moon were built in the south, north, east and west, respectively, of the Beijing city. In addition, the Jin imperial government also established the Altar of Land and Grain, and the Imperial Ancestral Temple for worshipping ancestors, as well as Yuan Temple (the Original Temple) for keeping emperors' portraits and the objects that emperors used before death.

### (2) The altars and temples in Beijing in the Yuan Dynasty

In the 9th Dade Year of the reign of Emperor Chengzong of the Yuan Dynasty (1305 A.D.), a temple was formally established to jointly worship Gods of Heaven and Earth, as well as the Five Prehistoric Emperors. In the 30th year of the Zhiyuan reign(1293 A. D.), the Altar of Land and Grain was built in a little south of the Inner Heyi Gate. The Imperial Ancestral Temple was ordered to be constructed in 1267, and was completed in the 17th year of the Zhiyuan reign(1280 A.D.). The Temple of Confucius was also established in Dadu,and remained in use till the Ming and Qing Dynasties.

### (3) The formation of the altar and temple patterns in Beijing in the Ming Dynasty

After Zhu Di, Emperor Chengzu of the Ming Dynasty, went into power in 1403, he decided to move the capital to Beijing. The altar and temple construction was an important part of the new capital. The construction of the altars and temples in suburbs started in June of the 15th year of the reign of Emperor Yongle of the Ming Dynasty (1417 A.D.), and the Temple of Heaven and Earth was completed in the 18th year of the reign of Emperor Yongle (1420 A.D.). Emperor Chenzu relocated the national capital from Nanjing to Beijing in the same year. During Emperor Chengzu's reign, Beijing's temple pattern formed based on that in Nanjing was such that the Imperial Ancestral Temple and the Altar of Land and Grain were situated on the left and right sides of the south of the Forbidden City, while the Temple of Heaven and Earth as well as the Temple of Mountain and River for offering sacrifice to the gods of heaven, earth, and numerous natural gods, were located on the left and right sides of the south of Zhengyangmen. The pattern remained in use for more than 160 years until the reign of Emperor Jiajing of the Ming Dynasty when the sacrifice rites were broken.

In the 9th year of the Jiajing reign (1530 A.D.), Emperor Zhu Houcong decided to restore the original system of Emperor Mingtaizu, i.e. separately worship to heaven and earth. Thus, the Circular Mound Temple (the Temple of Heaven), Fangze Temple (the Altar of Earth), the Zhaori Temple, the Xiyue Temple were located in the south, north, east, and west, respectively, of the city, worshipping heaven, earth, sun, and moon, respectively. The Circular Mound Altar was built to the south of the Grand Sacrificial Hall in the Temple of Heaven. In the 32nd year of the reign of Emperor Jiajing of the Ming Dynasty (1553 A.D.), the outer city wall in the south of

Beijing was completed for construction, the Temple of Heaven which was originally in the southern suburb, was thus included in the city of Beijing. On the southern end of the central axis of Beijing city was the Yongdingmen. The Temple of Heaven stands in the east of the southern section of Beijing's central axis, i.e. the line from the Yongdingmen of outer city to the Zhengyangmen of inner city, while the Imperial Ancestral Temple lies in the west of that section of the Beijing central axis. The Temple of Heaven was added with the outer wall with only one opening to the west, and the whole temple area was thus expanded to 2,730,000 square meters. However, the axis linking the northern and southern temples was in slight east of the central axis rather than in the middle, breaking the practice of building the major architectures on the central axis.

## (4) Renovation and design finalization of altar and temple architecture in the Qing Dynasty

In the 1st year of the reign of Emperor Shunzhi (1644 A.D.), the Qing Dynasty sent the envoy to worship the Temple of Heaven immediately after they entered the pass to Beijing. The temples built in the Ming Dynasty remained to be used during the reigns of Emperors Shunzhi, Kangxi, and Yongzheng in early Qing Dynasty.

In the 8th year of the reign of Emperor Qianlong (1743 A.D.), the Hall of Abstinence in the Temple of Heaven was started for renovation. In 1747, the seriously-damaged inner and outer walls was dismantled, and the Songyu Altar was approved to be demolished in the same year, and the rite for praying for rain was held in the Circular Mound Altar instead, thus representing the beginning of large-scale construction, renovation, and expansion of the altar and temple architecture during the reign of Emperor Gaozong in Qing Dynasty. In the 16th year of the reign of Emperor Qianlong (1751 A.D.), the former inscription, "Daxiang Hall", for the Hall of Prayer for Good Harvests was changed to "the Hall of Prayer for Good Harvests",

with meaning of praying for abundant harvests. Meanwhile, the Daxiang Gate was renamed Qinian Gate. For the renovation of the Hall of Prayer for Good Harvests, the color of glazed tiles was changed to purely blue from three colors, i.e. blue, yellow, and green, which symbolized the Heaven, the Emperor, and commoners respectively. The color alteration indicated the transformation from joint worship to the Heaven and Earth to purely pray for good harvests to the God of Heaven. The color of the tiles of the wing rooms of the Hall of Prayer for Good Harvests was changed from green to blue, while the three enclosing walls and gate-connecting walls of the Hall of Prayer for Good Harvests remained the former green tiles, due to its relatively far distance from the Temple of Heaven. In the 19th year of the reign of Emperor Qianlong (1754 A.D.), a gate, called the Gate of the Circular Mound Altar, was built at the south of the outer wall at the western gate of the Temple of Heaven, while the former western gate was called the Gate of Prayer for Harvests. The Temple of Heaven was thus divided into the northern and southern parts, presenting a precise pattern which has been remained till today. During the reign of Emperor Qianlong, great changes occurred in the altars and temples in Beijing, such as more brilliant colors, more sanctified and lofty architecture, the planting of larger quantity of woods, making the altars and temples more solemn and lucid.

In 1900, the imperial altars and temples were plundered by the Eight-Power Allied Forces after they conquered Beijing, furnishings, sacrificial utensils, ritual articles, and some other items kept in altars were all looted. In 1911, the Qing Dynasty was overthrown by the Xinhai Revolution, and the sacrificial rite system was subsequently rescinded. The altars and temples in Beijing was then under the jurisdiction of the Etiquette and Custom Department, Internal Affairs Ministry of the Government of the Republic of China. After being taken over, they were open successively as the public parks.

Nowadays, the majority of the altars and temples still

serve as parks or museums open to the public, enabling the masses to appreciate the charm of former-day royal altars and temples.

## 2. The Profile of the Altars and Temples Existing in Beijing

The altars and temples existing in Beijing, from the imperial altars and temples to celebrity ancestral temples, basically follow the pattern formed in the Ming and Qing Dynasties. They can be classified into the following categories:

(1) The altars of gods of heaven: used for worshipping natural gods of heaven, sun, moon, cloud, rain, thunder etc. These altars and temples have been well preserved, and most of which are imperial altars and temples with a spacious land area and higher architectural level.

(2) The altars of gods of earth: used for paying homage to the gods of earth, ancestral farm, ancestral silkworm, mountain, town, sea, river, famous mountain and large river and so on.

(3) Ancestral temples: there are numerous ancestral temples in Beijing, such as the Imperial Ancestral Temple where the emperors paid homage to their ancestors, the Temple of Successive Emperors, the Temple of Confucius, the temples for worshipping senior officials of virtue, e.g. the Temple of Wen Tianxiang, the Temple of Yu Qian, and the Temple of Yuan Chonghuan, as well as familial memorial halls. Compared with the first two types of temples, the familial memorial halls usually cover smaller area, and most of which have one or two courtyards or even just a house.

## 3. The Architecture and Art Achievement in the Altars and Temples in Beijing

Beijing is the only ancient oriental capital in the world with large-scale altar and temple architecture.

First, in terms of architectural technology, the ancient altars and temples existing in Beijing are the last masterpieces of Chinese ancient society, with the introduction of mature traditional techniques of all kinds. For example, the acoustics theory was successfully applied in the Echo Wall of the Imperial Vault of Heaven, as well as the Heaven's Heart Stone on the Circular Mound Altar in the Temple of Heaven. You will feel your voice more stentorian on the Heaven's Heart Stone.

Second, in terms of architectural art:

(1) Successful integration of the ancient universe outlook, ancient philosophy with architecture, makes the architecture reflect Chinese ancient philosophy and cosmography. For example, the four large pillars in the center of the Hall of Prayer for Good Harvests are known as the Dragon Well Pillars with each pillar representing one of the four seasons. The other 24 pillars are arranged in two circles surrounding the four central ones. The 12 inner pillars represent the 12 months of the year and the 12 outer pillars represent the divisions of day and night. The integration of the inner and outer pillars represents 24 Jieqi (a terminology used in the calendar to track the seasonal changes throughout a year). The total 28 pillars represent the 28 mansions in the heaven. Such symbolic representation means used in all temple architecture in Beijing reflects the connotation of Chinese traditional culture.

(2) The application of artistic conception in architecture reaches its peak in Chinese traditional buildings. The view-presenting methods, such as repression, exaltation, suspension, transition, and duplication, as well as various techniques such as matching the high and low buildings, and color contract, are used incisively and vividly, creating an atmosphere of tranquility, vastness, and solemness. For example, the altars and temples are surrounded by luxuriant cypresses. If looking around from the high point in the temple, one will have an eyeful of greenness, and feel as if he was in the universe with high sky and boundless earth, and

spontaneously worship the heaven, earth, sun and moon.

(3) The adoption of architectural language. The architectural language adopted in Beijing is mainly that, as said in the Yin Yang Theory (a learning that studies the negative and positive features and changes of both nature and human beings), Heaven is "Yang", and Earth is "Yin". Therefore, the site of the Temple of Heaven was chose in the south, namely the Yang position, while the Altar of earth was located in the north, i.e. the Yin position. The metaphorical or symbolic numbers were systematically applied in the details of the architecture, e.g. the numbers one, three, five, seven, nine, which belong to the Yang category, were used in the Temple of Heaven, while the number two, four, six, eight, which belong to the Yin category, was used in the Altar of Earth. According to the theory of "Round Heaven and Square Earth", the Circular Mound Altar and the Hall of Prayer for Good Harvests,

and other sacrificial architectures are all circular in shape to symbolize the round heaven, while the Fangze Altar and other sacrificial architecture are all square-shaped to symbolize the square earth. The walls of the Temple of Heaven are square-shaped in the south side and round-shaped in the north side. The metaphoric meaning of Round Heaven and Square Earth also is represented in round-shaped inner parapet wall and square-shaped outer parapet wall of the Circular Mound Altar in the Temple of Heaven, in the square outer parapet wall of the Fangze Altar, the circular parapet wall of the worshiping terrace of the Altar of the Sun and the square-shaped parapet wall of the Altar of the Moon.

Staying in these ancient altars in Beijing, you will feel their profound history and unique structures and have infinite mental enjoyment.

# 天坛
# The Temple of Heaven

  天坛位于北京市永定门内大街东侧，是明清两代皇帝祭天和祈求五谷丰收的地方，是现存中国古代王朝等级最高、最完整、最有特色的坛庙建筑群之一。1961年被公布为全国重点文物保护单位，1998年被列入世界文化遗产名录。

  天坛始建于明永乐十八年（1420年），定址在正阳门南3.5公里，北京城中轴线稍偏东的位置，合祀天地，名天地坛，主体建筑为矩形的大祀殿。嘉靖九年（1530年）在北郊建地坛，天地分祀，此处专门祭天，改称天坛，并在大祀殿南建专门行祭天礼的圜丘坛，嘉靖十九年（1540年）又在大祀殿建行祈谷礼的大享殿，至此天坛的形制大致形成了。嘉靖二十四年（1545年）祈年殿改为三重檐攒尖顶，殿顶覆盖上青、中黄、下绿三色琉璃瓦，寓意天、地、万物。清乾隆年间，改圜丘为蓝琉璃栏杆、地面砖为石制，改皇穹宇的重檐顶为单檐顶，将祈年殿三层檐的蓝、黄、绿三色琉璃瓦一律改为蓝琉璃瓦，成为现在天坛的面貌。1889年祈年殿毁于雷火，1890年按照原式重建。

  天坛占地面积约273万平方米，由两重坛墙分为内坛和外坛两部分，坛墙南为方形北为圆形，以象征中国古代认为的"天圆地方"，外坛墙总长6553米，内坛墙总长4152米，主要建筑物有祭天的圜丘和祈谷的祈年殿，疏朗地布置在稍稍偏东的天坛南北中轴线的两端，中央以一条称为"丹陛桥"的高甬道连接。还有一些附属建筑，最大的是内坛的斋宫和外坛内的神乐署。

  天坛从选位、规划布局、建筑的设计以及祭祀礼仪和祭祀乐舞，都有深刻的文化内涵，它成功地把古人对"天"的认识、"天人关系"以及对来年的美好祈愿用建筑的形式表现出来，是一座集古代哲学、历史、数学、力学、美学、生态学、建筑学、风水术等于一身的古代建筑精品，建筑处处体现了中国古代特有的象征和寓意，是中国现存又一座集大成的古建筑群，在世界上享有盛誉。

Located to the east of Yongdingmennei Street, the Temple of Heaven was the sacred place where the emperors of the Ming and Qing Dynasties worshipped heaven and prayed for good harvests. It is one of the largest and the most complete sacrificial temples in China. Listed as a national key relic under special preservation in 1961, the Temple of Heaven was also inscribed in the World Heritage List of UNESCO in 1998.

About 3.5 kilometers to the south of the South-Facing Gate the temple was first built in the 18th year of the reign of Emperor Yongle in the Ming Dynasty (1420 A.D.). The main building was the Grand Sacrificial Hall shaped like

鸟瞰天坛主轴线
The Bird View of the Central
Axis of the Temple of Heaven

a rectangle. It was built as the Temple of Heaven and Earth, but was given its current name in the 9th year of the reign of Emperor Jiajing (1530 A.D.), who built the Altar of Earth in the northern suburbs. In 1540, the original Grand Sacrificial Hall was no use and rebuilt into the Daxiang Hall, namely, the Hall of Prayer for Good Harvests, which was changed into a triple-eave pyramidal roof covered with green, yellow and blue glazed tiles from the first eave to the third, meaning everything, earth and heaven respectively in 1545. During the reign of Emperor Qianlong, the roof of the Hall of Prayer for Good Harvests was changed into blue glazed tiles, following the colour of the sky.

The Temple of Heaven occupies about 2,730,000 square meters of land and is divided into the outer altar and the inner altar by two enclosed walls. The section of wall enclosing the southern end of the temple grounds is square, while that in the northern end is semi-circular, based on the ancient notion that the Earth is square and Heaven round. The circumference of the outer altar wall is 6,553 meters and of the inner altar wall is 4,152 meters. The Bridge of Cinnabar Steps connects the Hall of Prayer for Good Harvests to the north and the Circular Mound Altar to the south.

圜丘
The Circular Mound Altar

内坛的圜丘是皇帝举行祭天大典的地方。圜丘建筑充满了象征意义。外面有二重围墙，内圆外方，象征"天圆地方"。两重墙四面各开辟白石建造的棂星门，围墙中间是一座三层圆形石坛台。每层台各出九级台级，上层台面中心一块圆形石板称"天心石"，匠师们利用声音反射原理，使人站在天心石上说话，声音特别浑厚、洪亮，如同声音直达天庭，是皇帝祭天时宣读祭文所立之处。围绕天心石四周环砌扇形台面石，第一圈9块，依次往外每圈递增9块，直至"九九"81块，以寓意我国古代认为的"九重天"，中下层也都是9的倍数。栏板三层共360块，象征周天360度。坛的尺寸也是有寓意，尺寸按照古尺设计，上层直径9丈，取一、九数；中层15丈，取三、五数，下层直径21丈，取三、七之数，加在一起"一、三、五、七、九"单数中5个"阳数"全部包括。

圜丘棂星门
The Star-Worshipping Gate

天心石
The Stone of Heaven's Center

皇穹宇院建筑
The Hall of the Imperial Heavenly Vault

圜丘往北为皇穹宇，是用于平日供奉祀天大典时所供神版的殿宇。皇穹宇殿
为单檐圆攒尖顶，蓝琉璃瓦，鎏金宝顶，高19.5米，直径15.6米。殿四周环
以充分利用声学原理的圆形围墙，声波可由墙壁传递，即著名的"回音壁"。

皇穹宇内景
Interior of the Hall of the Imperial Heavenly Vault

皇穹宇殿内有8根金柱，均为香楠木，尚为明代原物，柱身满绘红地
金花的缠枝莲，保存着明代彩画的特点，殿顶为青绿基调的金龙藻
井，中心为大金团龙图案，显得严谨、精致，殿内供奉"皇天上帝"神
牌位，配祀清代8位祖先牌位。

丹陛桥
The Bridge of Cinnabar Steps

皇穹宇往北有祈年殿、皇乾殿，由一条高3.35米、宽29.4米、长359米
的高甬道把这两组建筑连接起来，即"丹陛桥"，又称"海墁大道"。甬
道向北微微高起，似步步高升，站在丹陛桥上环望四周，地下无尽的
苍松翠柏，满眼翠碧，似近天宇。

祈年门匾
The Board Inscribed with "the Gate of
Prayer for Good Harvests"

由祈年门内望祈年殿
Looking at the Hall of Prayer for Good Harvests from the Gate of Prayer for Good Harvests

祈年殿正立面图
The Elevation of the Hall of
Prayer for Good Harvests

祈年殿
The Hall of Prayer for Good Harvests

祈年殿是孟春(正月)皇帝祈祷丰年的专用建筑，也是天坛中体量最大的建筑物，又是天坛建筑群的中心，它建在一个与丹陛桥同高的砖砌台子上，台子东西宽约165米，南北长约191米，台顶边沿建有一道高两米的矮墙围成庭院，庭中央偏北中轴线上为一座三层的圆形坛台，名"祈谷坛"，每层各出八级台级，四周均以石栏围护。坛中央建祈年殿。殿高38米，为三重檐圆攒尖顶，向上层层收缩直冲霄汉，上覆蓝琉璃瓦，鎏金宝顶。

祈年殿院落空间开朗，主体建筑高大而又有特色，外形轮廓上由逐层收缩的坛台、屋檐和圆锥形屋顶造成强烈的向上的感觉，内部也层层增高，聚拢向中间，营造出与天接近的气氛，站在坛台上向四周环望，两侧的庑房与背后的皇乾殿的屋顶与台面相平，越过矮墙还可以看到柏林的树梢，仿佛漂浮在一片绿海之上，所有这一切构成了祈年殿整体感觉的恢宏和神秘。

祈年殿宝顶
The Spire of the Hall of
Prayer for Good Harvests

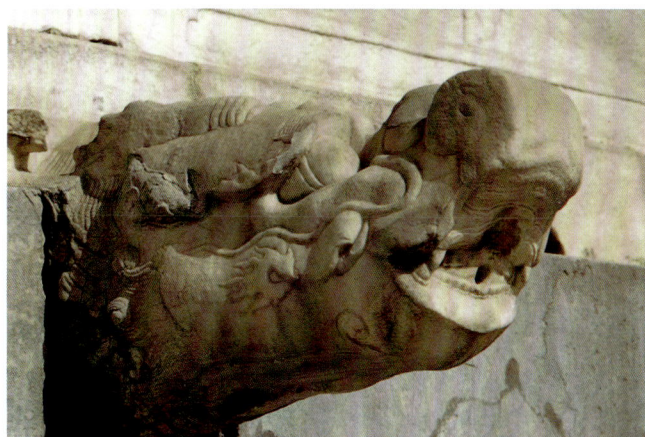

祈年殿汉白玉雕出水——龙
The Waterspout of the Stylobate of the Hall of
Prayer for Good Harvests: Dragon

祈年殿汉白玉雕出水——凤
The Waterspout of the Stylobate of the Hall of
Prayer for Good Harvests: Phoenix

祈年殿内景
Interior of the Hall of Prayer for Good Harvests

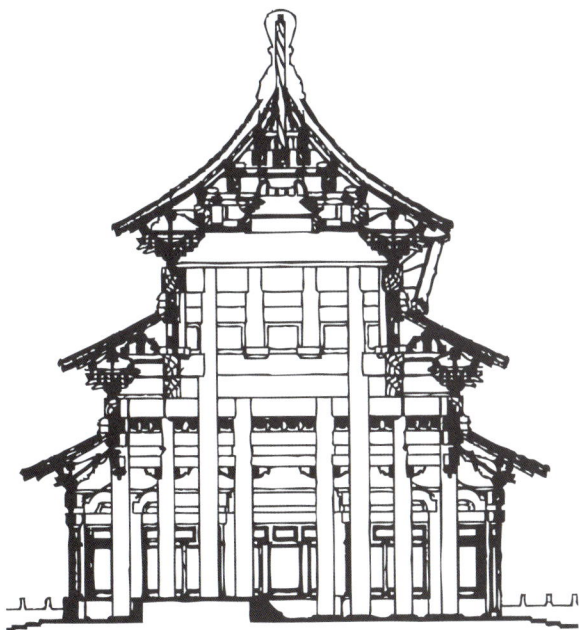

祈年殿内部结构图
Cross Section of the Hall of Prayer for Good Harvests

祈年殿殿内有楠木大柱28根环绕排列，里圈的4根龙井柱寓春夏秋冬四季，中间一圈12根金柱寓一年12个月，最外一圈12根檐柱寓意一天12时辰，中圈和外圈相加寓一年二十四节气，内外总计28根寓意周天二十八星宿。殿顶绘金龙藻井，殿中央地面上有一块天然龙凤纹大理石，殿中雕龙宝座(祭祀时放神牌位用)，加上龙凤和玺彩画交相辉映，显得金碧辉煌。

皇乾殿
The Hall of Imperial Heaven

长廊
Long Corridor

斋宫全貌
The Palace of Abstinence

斋宫位于内坛西南隅，是皇帝到天坛祭祀时"斋戒"的地方。皇帝每逢祭天典礼之前便住进斋宫，这期间皇帝声色、荤腥俱禁止沾染，以表示对天的尊敬。斋宫坐西朝东，平面正方形，面积近4万平方米，四周以两重城壕和两重宫墙环护，建筑瓦用绿色以表达帝王向天称臣、尊敬恭谦之意。主要建筑有无梁殿、寝殿，另外还有斋戒铜人石亭、钟鼓楼、值守房和巡守步廊等礼仪、居住、服务、警卫专用建筑。

斋宫二宫门
The Second Gate of the Palace
of Abstinence

斋宫内的无梁殿
The Beamless Hall

铜人亭
The Pavilion of Bronze Figure

神乐署凝禧殿
The Ningxi Hall of the Divine Music Hall

神乐署是外坛的主要建筑，位于外坛西门内稍南侧，坐西朝东，是天坛五组大型建筑之一，是专司明清两代皇家祭祀大典乐舞的机构。神乐署建于明永乐十八年(1420年)，称作神乐观，是一所专司祭祀音乐的道观。清乾隆年间改为神乐署，当时京城各个皇家祭坛的祭祀乐舞生皆由天坛神乐署生员中选拔充任。清末被八国联军侵占，神乐署从此衰败。神乐署建筑总平面呈东西长、南北短的长方形，为两重殿宇的三进院落。神乐署大门朝东，前殿五开间，明称太和殿，清康熙年间改名为凝禧殿，用于排演祭祀大典；后殿七开间，原名玄武殿，明末改称显佑殿，用于供奉玄武大帝以及诸乐神；殿后还有袍服库、典礼署、奉祀堂等建筑，东庑由通赞房、恪恭堂、正伦堂、候公堂、穆佾所等建筑，西庑有掌乐堂、协律堂、教师房、伶伦堂、昭佾所等建筑。

# 地坛
# The Altar of Earth

地坛位于北京市东城区安定门外大街东侧，明清北京五大坛之一，是明清两代皇帝"夏至"日祭祀地皇神祇的地方。1984年被公布为北京市文物保护单位。

地坛始建于明嘉靖九年（1530年），初名方泽坛，嘉靖十三年（1534年），改名地坛。明万历，清雍正、乾隆、嘉庆、同治年间屡加扩建、修缮。

地坛占地面积42.7公顷，是仅次于天坛的第二大祭坛。地坛分为内坛和外坛，以祭祀为中心，周围建有皇祇室、斋宫、神库、神厨、宰牲亭、钟楼等。举行祭地大典的方泽坛是最重要的建筑，平面为正方形，二层，上层高1.28米，边长20.5米，下层高1.25米，边长35米。地坛的建筑处处体现出我国古代对"地"的认识，是北京最优美的建筑群之一。

The Altar of Earth is located on the east of Andingmenwai Street. Being one of five altars in the Ming and Qing Dynasties, it was the sacred place where the emperors of the Ming and Qing Dynasties worshipped the God of Earth at the Summer Solstice and was listed as a Beijing's relic under preservation in 1984.

Construction of the altar began in the 9th year of the reign of Emperor Jiajing of the Ming Dynasty (1530 A. D.), known as Fangze Altar. The altar was renamed the Altar of Earth in 1534 and was repeatedly enlarged and reconstructed in the Ming and Qing Dynasties.

The Altar of Earth occupies about 42.7 hectares of land and comprises the outer altar and the inner altar. The principal structures are the Palace of Abstinence, the Divine Storehouse, the Divine Kitchen, the Sacrificial Animals Pavilion, the Bell Tower and so on.

方泽坛
The Fangze Altar

# 日坛
# The Altar of the Sun

日坛又名朝日坛，位于北京市朝阳区朝阳门外日坛路东，明清北京五大坛之一，是明清两代皇帝"春分"节祭祀太阳神的地方。1984年被公布为第三批北京市文物保护单位。

日坛始建于明嘉靖九年（1530年），清代曾几次修缮。日坛主要建筑有拜坛、具服殿、神厨、神库、钟楼、祭器库等。日坛建筑也同样体现了象征意义，表达了我国古代对"日"的认识，是北京坛庙建筑的又一座精品。

Located on the east of Ritan Street outside the city gate of Chaoyangmen in Chaoyang District, the Altar of the Sun also known as the Altar of the Rising Sun, was one of the five altars in the Ming and Qing Dynasties. The altar was the sacred place where the emperors of the Ming and Qing Dynasties worshipped the God of Sun at the Vernal Equinox and was listed as a Beijing's relic under preservation in 1984.

First built in the 9th year of the reign of Emperor Jiajing of the Ming Dynasty (1530 A.D.), it was renovated several times in the Qing Dynasty. The main buildings include the altar, the Hall of Changing Clothes, the Divine Kitchen, the Divine Storehouse, the Bell Tower and so on.

具服殿木影壁
The Wooden Screen Wall of the
Hall of Changing Clothes

日坛
The Altar of the Sun

# 月坛
## The Altar of the Moon

　　月坛位于北京市西城区月坛北路南，又名夕月坛，明清北京五大坛之一，是明清两代皇帝"秋分"祭祀夜明之神（月亮）和天上诸星宿的地方。2006年被公布为全国重点文物保护单位。

　　月坛始建于明嘉靖九年（1530年）。月坛主要建筑有拜坛、具服殿、神厨、神库、钟楼等。月坛虽然占地面积较天、地坛小很多，但是祭坛所有的建筑基本上都具备了，而且其建筑也是处处都体现了象征意义，表达了中国古代对月神的认识。

Located on the south of North Yuetan Street in Xicheng District, the Altar of the Moon also known as Xiyue Altar was one of the five altars in the Ming and Qing Dynasties. The altar was the sacred place where the emperors of the Ming and Qing Dynasties worshipped the God of Moon and mansions at the Autumnal Equinox and was listed as a national key relic under special preservation in 2006.

First built in the 9th year of the reign of Emperor Jiajing of the Ming Dynasty (1530 A.D.), the main buildings include the altar, the Hall of Changing Clothes, the Divine Kitchen, the Divine Storehouse, the Bell Tower and so on.

钟楼
The Bell Tower

北天门
The Gate of the Northern Heaven

月坛棂星门
The Star-Worshipping Gate of the Altar of the Moon

# 社稷坛
# The Alter of Land and Grain

　　社稷坛位于北京市东城区西长安街北侧、天安门之西，是明、清两朝皇帝每年春秋仲月上戊日祭祀社（土地神）、稷（五谷神）的场所。1988年被公布为全国重点文物保护单位。

　　明永乐十八年(1420年)，成祖朱棣迁都北京，根据"左祖右社"的帝王都城设计原则，在紫禁城的前右侧兴建社稷坛，清因明制，这里依然是清代的社稷坛。清帝逊位后，民国三年(1914年)，社稷坛辟为公园，时称中央公园；陆续增建了一些风景建筑和纪念建筑，如唐花坞、投壶亭、春明馆、绘影楼、长廊等。民国十七年(1928年)为纪念孙中山先生，改称为中山公园。

　　社稷坛的平面为一南北稍长的不规则长方形，南部东西宽345.5米，北部东西宽375.1米，南北长470.3米，总面积约为24公顷。园内的主体建筑有社稷坛、拜殿、戟门等；还有一些辅助建筑，如宰牲亭、神库和神厨等。

社稷坛拜殿
The Worshipping Hall of the Alter of Land and Grain

Located to the north of Xichang'an Street and to the west of the Gate of Heavenly Peace in Dongcheng District, the Alter of Land and Grain was the sacred place where the emperors of the Ming and Qing Dynasties worshipped the gods of land and grain. The altar was listed as a national key relic under special preservation in 1988.

According to the guiding principle for planning the royal city that the Imperial Ancestral Temple should be located on the left while the Alter of Land and Grain on the right, the altar stands on the right front of the Forbidden City. First built in the 18th year of the reign of Emperor Yongle of the Ming Dynasty (1420 A.D.), it shaped like an irregular rectangle covers an area of 24 hectares.

五色土
Five Colors of Soil

# 先农坛
# The Altar of the God of Agriculture

先农坛位于北京中轴线南端的永定门内大街西侧，与天坛隔街东西呼应，又名山川坛，是明清两朝皇帝祭祀先农、天神地祇、太岁诸神及举行耕籍典礼的场所。2001年被公布为全国重点文物保护单位。

先农坛始建于明永乐十八年(1420年)，由内外两重坛墙环绕，围墙平面北圆南方，总面积130公顷。外坛墙已于民国初年被拆除，现存古建筑群有太岁殿、神仓、庆成宫、神厨库、先农坛神坛、宰牲亭、拜殿、具服殿、观耕台等，是明清皇家祭祀建筑中保留较为完整的一处。

Located to the west of Yongdingmennei Street, the Altar of the God of Agriculture is directly to the west of the Temple of Heaven. The altar was the sacred place where the emperors of the Ming and Qing Dynasties worshipped Shennong, a legendary farmer of great antiquity. It was listed as a national key relic under special preservation in 2001.

First built in the 18th year of the reign of Emperor Yongle of the Ming Dynasty (1420 A.D.), the altar is surrounded by two enclosed walls, covering an area of 130 hectares. The existing ancient buildings complex includes the Hall of the Year God, the Divine Storehouse, the Palace of Celebrating Completion, the Divine Kitchen, the altar, the Sacrificial Animals Pavilion, the Hall of Changing Clothes and so on.

神厨院正殿
The Main Hall of the Divine Kitchen Courtyard

神厨院位于太岁殿西侧，占地面积6530平方米，为存放先农诸神神位和准备牺牲祭品的场所。正殿供奉先农诸神牌位。西殿——神厨为祀前准备祭品的地方，建筑后檐窗外有一汉白玉水槽。东殿——神库为存放祭祀用具的地方。两座井亭为打理祭祀品时的取水之处，为六角形，盝顶屋面，中心与井口上下相对，寓意天地一气。宰牲亭位于神厨西侧，是祭祀先农坛内诸神时宰杀牺牲的场所，其屋顶采用悬山重檐的形式，被誉为中国古建筑中的孤品。

太岁殿
The Hall of the Year God

太岁殿始建于明永乐十八年(1420年)，是祭祀太岁神及十二月将神的场所。太岁殿面阔七间，黑琉璃筒瓦绿剪边歇山顶，专祀值年之神——太岁神；东西配殿各十一间，分祀四季和各六位月将神；南殿七间，为拜殿。清乾隆初年对太岁殿院落进行了大修。

拜殿
The Worshipping Hall

观耕台是皇帝亲耕完毕后观看王公大臣耕作之处。初建于明嘉靖十年(1531年)，原为木构，每年亲耕时临时搭建。清乾隆十八年(1753年)改建为砖石结构，台呈方形，边长16米，高1.5米，东、南、西三面各出8阶，黄绿色琉璃台基饰浮雕，上置汉白玉栏杆。一亩三分地为皇帝亲耕礼的藉田，位于观耕台正南。

观耕台
The GuanGeng Platform

具服殿
The Hall of Changing Clothes

具服殿始建于明永乐十八年(1420年)，嘉靖十一年(1532年)为皇帝亲耕前更换亲耕礼服的场所。

神坛
The Altar

位于先农坛内坛西北，原为皇帝露祭先农神的祭坛。砖石结构，四面各出八级台阶。明清时，每年仲春或季春吉亥之日，皇帝亲临或遣官在此祭先农，随后藉田行耕藉礼。

神祇坛在先农坛内垣外西南，四周有红墙围绕，原坛墙南北各有一座三座门，现仅有南门残存。墙内有两座坛，东为天神坛，西为地祇坛，均建于明嘉靖十一年（1532年）。天神坛，南向，砖石结构，方形，面积17平方米，四面墙墙，并有座石雕棂星门，坛北有白石龛4座，上雕流云、海水文饰，分别祭祀风、云、雷、雨四神。现仅存南墙上的三门六柱棂星门及一座焚帛炉。地祇坛，南向，砖石结构，方形，面积33平方米，四面墙墙及石雕棂星门，坛南有青白石龛五座，分祀五岳五镇五山四海渎神，京畿山川神及天下山川神。现在坛已无存，仅剩北墙三门六柱棂星门，东西二墙墙各一门二柱棂星门及南墙的三门六柱棂星门；还有残缺的神位石龛，为了更好地保护，于2002年将其移至太岁殿院西南安置。

原神祇坛石龛
Niches of the Former Shenzhi Altar

庆成宫前殿
The Front Hall of the Palace of Celebrating Completion

庆成宫建筑群位于内坛东门外迤北，占地约13000平方米，建于明天顺二年（1458年），原称斋宫。清乾隆二十年（1755年）后改称庆成宫，为皇帝祭祀先农行耕礼后、犒劳随从百官之所。庆成宫设宫门两重，东南墙内侧原有钟楼一座，院内主要建筑有前殿、后殿及东西配殿。

神仓院
The Divine Storehouse Courtyard

神仓院位于先农坛东北，占地3500平方米，分为南北二进院落，用于
贮藏耕耤田收获的谷物，以备京城各坛庙祭祀所需。前院有山门、收
谷亭、圆廪神仓、两侧为粮仓、碾房，后院为祭器库。

神仓碾房
The Grinding Hall of the Divine
Storehouse

先农坛神仓圆廪
Yuanlin of the Divine Storehous

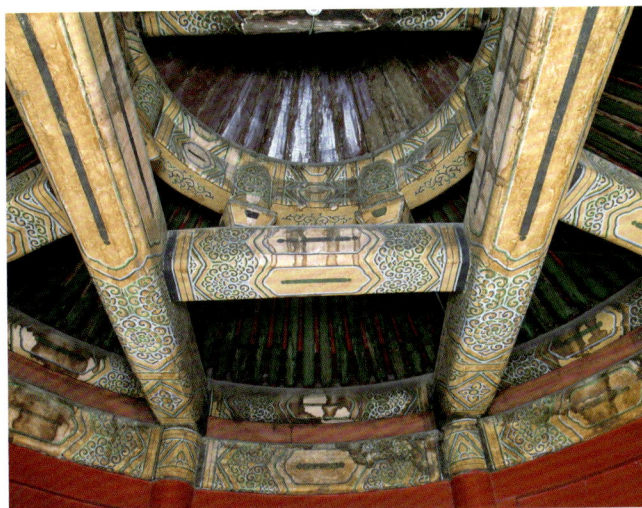

圆廪梁架彩画
Coloured Paintings at the
Framework of Yuanlin

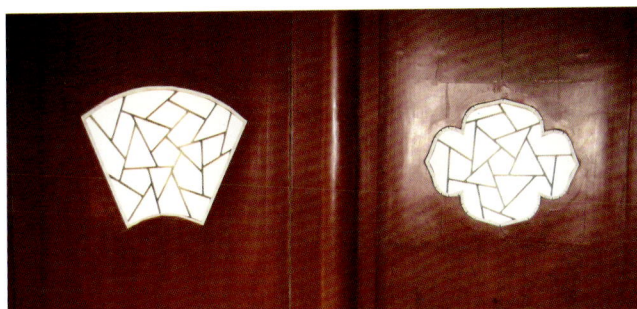

圆廪什锦窗
Windows of Various Forms of
Yuanlin

# 凝和庙
# Ninghe Temple

凝和庙位于北京市东城区北池子大街46号，是奉祀云神之处。1984年被公布为北京市级文物保护单位。

凝和庙俗称云神庙，清雍正八年（1730年）敕建，仿照宣仁庙规制建造。由于其地离紫禁城近，所以旧时常有进京述职、办事的官员居住在此。民国时，此处改为学校，现为北池子小学。

庙门坐东朝西，面阔三间，单檐歇山顶，黄琉璃瓦绿剪边。门前有一座琉璃砖砌大影壁，建于石须弥座上，现已无存。庙内殿宇均坐北朝南，主要建筑有钟鼓楼及四重大殿。钟鼓楼二层，平面方形，歇山顶，黄琉璃瓦绿剪边。献殿三间，硬山调大脊，黑琉璃瓦绿剪边，木构架绘旋子彩画；正殿三间，单檐歇山顶，黄琉璃瓦绿剪边。后殿五间，单檐歇山顶，黄琉璃瓦绿剪边。后殿东西两侧各有朵殿三间，硬山过垄脊筒瓦屋面。目前，钟鼓、献殿等已拆除，只有山门、大殿及殿前御道、后殿还保留。

Located at No.46 Beichizi Street in Dongcheng District, the Ninghe Temple is commonly known as Yunshen Temple. Built in the 8th year of the reign of Emperor Yongzheng of the Qing Dynasty(1730 A.D.), it was listed as a Beijing's relic under preservation in 1984.

Facing west, the front gate is 3 bays wide, with a gable-and-hip roof. The roof is covered with yellow glazed tiles with a green edge. The main hall is 3 bays wide with a-gable-and-hip roof. The roof is covered with yellow glazed tiles with a green edge. The rear hall is 5 bays wide with a gable-and hip roof. The roof is covered with yellow glazed tiles with a green edge. Now, only the front gate, the main hall, the imperial road and the rear hall remain.

山门
The Front Gate

大殿
The Main Hall

天花
The Ceiling

# 太庙
# The Imperial Ancestral Temple

太庙位于东长安街北侧，天安门城楼之东，是明、清两朝帝王的祖庙。1988年被公布为全国重点文物保护单位。

太庙始建于明永乐十八年（1420年）。太庙自建成后，历经多次修缮，明嘉靖、万历年间多次重修，规模均较大。明末被李自成起义军焚毁一部分建筑，清顺治年间重修。乾隆中晚年间，又对主体建筑大规模扩建。虽经多次修缮改建，其规模和建筑却大体保持原状，成为北京市最完整的明代建群之一。

太庙建筑布局呈南北向的长方形，总面积为139650平方米，内外共有三道黄琉璃瓦顶红墙身的围墙。最外一道长475米、宽294米。西侧辟有西向大门两座：南边一座称太庙街门，可通天安门里；北边一座叫太庙庙门，可达端门里。外围墙内是太庙的第一层院落，满植成排的古柏。此院东南角有一所西向的房院，是为太庙牺牲所。牺牲所的西侧有六角井亭。第二道围墙俯视平面亦呈长方形，东西宽208米，南北长272米。太庙的主要建筑均在这道围墙内，它们是戟门、前殿、中殿和后殿。

前殿的东西配庑各15间，形制相同：黄琉璃筒瓦歇山顶，殿式做法。东配庑前有大燎炉一座，为焚化前殿及东庑的祝帛之用。西配庑南侧有小燎炉一座，为焚化西庑的祝帛之用。

It is located to the north of Dongchang'an Street and to the east of the Gate of Heavenly Peace and once served as the imperial ancestral temple of the Ming and Qing Dynasties. It was listed as a national key relic under special preservation in 1988.

First built in the 18th year of the reign of Emperor Yongle of the Ming Dynasty (1420 A.D.), the temple was renovated several times during the Ming Dynasty. In the last years of the Ming Dynasty, some buildings were destroyed by Li Zicheng's army. During the reigns of Emperor Shunzhi and Qianlong in Qing Dynasty, it was renovated and enlarged.

The temple shaped like a rectangle covers an area of 139,650 square meters. It is all surrounded by three enclosed red walls, the roof of which is covered with yellow glazed tiles. The principle buildings consist of three main halls, two gates, two subsidiary shrines, and various accompanying buildings.

太庙匾额
The Board Inscribed with "the Imperial Ancestral Temple"

一进琉璃门
The Glazed Gate in the First Courtyard of the
Imperial Ancestral Temple

戟门
The Ji Gate

戟门面阔五间，正中三间为三座实踏大门，黄琉璃筒瓦庑殿顶，三层汉白玉石台基四
周都有石护栏；正中有汉白玉石雕御路，从上至下分别雕"二龙戏珠"、"狮滚绣球"、
"海水江涯"。东西旁门各一座，黄琉璃筒瓦歇山顶，无台基；面阔各一间。

前殿
The Main Hall

前殿在戟门正北，又称大殿，黄琉璃筒瓦重檐庑殿顶，殿式做法；面阔十一间，明间之上的两层檐间有匾额书"太庙"，满汉文竖写；头层檐下除尽间外均装六抹三交菱花格扇门或窗，尽间面阔很窄，装四抹三交菱花窗一扇。前殿梁柱均包镶沉香木，其余木构件为金丝楠木制成，地面墁铺金砖。殿基为汉白玉石须弥座，共三层，俗称"三台"。殿前有宽敞的月台，三层台基均有汉白玉石护栏，正中御路为三层，分别雕"云龙纹"、"狮滚绣球"和"海兽纹"。前殿是太庙"袷祭"的祭场，其内分昭穆设历代帝、后的神座，每代帝、后神位前摆一张笾豆案，案上摆放祭器和礼器，祀时，案前伴以佾舞。

寝殿
The Sleeping Hall

寝殿在前殿的后面，也叫中殿，黄琉璃筒瓦庑殿殿顶，殿式做法，面阔
九间，明间、次明间各装四抹三交菱花门4扇；梢间、尽间各装四抹
三交菱花窗4扇。台基为汉白玉石须弥座，殿前有月台，上绕以石护
栏。中殿内原供奉历代帝、后神龛，每龛外列放一代帝、后神椅、龛
内供奉一代帝、后神主牌位。祀飨时，先一日由官员上香，及期将神
牌置神椅上，移至前殿，奉安神座木托上。

太庙井亭
The Well Pavilion

云鹤柱头
Pillar Cap Carved with Crane and Cloud

转角处斗拱
Corbel Brackets on the Turning Part

# 孔庙
# The Temple of Confucius in Beijing

孔庙坐落在北京市东城区安定门内国子监街13号，与国子监相邻，是元、明、清三代皇帝祭祀孔子的场所。1988年被公布为全国重点文物保护单位。

孔庙始建元代。元世祖忽必烈定都北京后，下令袭历代旧典，命宣抚王楫于金枢密院建宣圣庙，祭祀孔子。到了元成宗铁穆耳大德六年(1302年)，在今址正式建庙，于大德十年(1306年)建成。大德十一年(1307年)特诏命孔子加谥为"大成至圣文宣王"，这块"加号诏书"石碑，现仍耸立在大成门前。元文宗至顺二年(1331年)，皇帝下诏恩准孔庙配享宫城规制，许孔庙四隅建角楼。元末，孔庙荒废。明永乐九年(1411年)，又重新整治，并修缮了大成殿。宣德四年(1429年)修整了大成殿及两庑。嘉靖九年(1530年)为祭祀孔子五代先祖增建崇圣祠。清乾隆二年(1737年)皇帝亲谕孔庙使用最高贵的黄琉璃瓦顶，只有崇圣祠仍用绿琉璃瓦顶。这时的孔庙已是红墙黄瓦，金碧辉煌了。光绪三十二年(1906年)祭孔的礼节升为大祀，孔庙也大规模地修缮。工程尚未完成，清朝被推翻，但修缮仍继续进行，到了民国五年(1916年)才最后竣工。至此孔庙形成了今天的规模和布局，成为仅次于曲阜孔庙的全国第二大孔庙。

北京孔庙占地面积22000平方米。院内以大成殿为中心，南北成一条中轴线，三进院落，左右建筑对称配列。轴线上建有嵌琉璃花砖一字影壁、先师门、大成门、大成殿等主要建筑。此外在大成门外东面设碑亭、省牲亭、井亭、神厨；西面有碑亭、致斋所、神库，并设有持敬门与国子监相通。最后一进是独立小院——崇圣祠。

The Temple of Confucius in Beijing's at No.13 Guozijian Street inside the city gate of Andingmen was the place where emperors offered sacrifices to Confucius during the Yuan, Ming and Qing Dynasties. It was listed as a national key relic under special preservation in 1988.

First built in the Yuan Dynasty, the Temple of Confucius was enlarged and repaired in the Ming Dynasty, and in the 2nd year of the reign of Emperor Qianlong of the Qing Dynasty (1737 A.D.), all the roofs of the halls were covered with yellow glazed tiles. Another large-scale reconstruction took place in the 32nd year of Emperor Guangxu (1906 A.D.) when Confucius was honoured with a grand memorial ceremony. Being repaired, the size of the temple is only smaller than the Temple of Confucius in Qufu.

The Temple of Confucius occupies about 22,000 square meters of land and comprises four courtyards one behind another. The principal structures, from the front to the rear, are the Ancient Teacher Gate, the Great Accomplishment Gate, the Great Accomplishment Hall and the Shrine for Honouring the Sage. On the eastern side of the front courtyard are the tablet pavilions, the Sacrificial Animals Pavilion, the Well Pavilion and the Divine Kitchen; on the western side are more tablet pavilions, the Vegetarian Dining Hall, the Divine Storehouse and the Respect-Paying Gate which leads to the Imperial Academy.

先师门
The Ancient Teacher Gate

孔庙鸟瞰
The Bird View of the Temple of Confucius

大成门
The Great Accomplishment Gate

大成门是崇基石栏，门前后三出
陛，中为丹陛，左右各13级。门内
悬钟置鼓各一，两侧放石鼓10枚。
大成门左右辟有角门。

大成殿
The Great Accomplishment Hall

大成殿坐落在以汉白玉雕云头石柱栏杆的月台上。月台左右砌石阶，两石阶中间嵌着一块7米长、2米宽的大青石浮雕，上下皆雕有飞龙戏珠，中间盘龙吞吐火焰宝珠，周围云水波涛。院内东西两边各有配庑19间，左右对称，布局规整。

大成殿是孔庙的主体建筑，是祭孔的正殿，始建于明永乐年间。光绪三十二年(1906年)将殿由七间三进、扩建为九间五进，瓦顶为重

檐庑殿顶，用光彩夺目的黄色琉璃瓦铺砌，殿顶正脊两端装饰着龙形鸱吻。

大成殿两侧有灰瓦通脊、单檐歇山顶式配庑。内部是放置从祀的历代先儒哲人牌位的地方。

大成殿内景
Interior of the Great Accomplishment Hall

殿内正中设木龛，龛内置"大成至圣文宣王"木牌位[明嘉靖年间、龛内曾设孔子泥塑像、后改为画像及木牌位，明成化十二年(1476年)曾特诏许孔子圣像穿戴帝王衣冠]，龛前置祭案、案上面摆设有尊、爵、卣，以及笾、豆等祭具。殿内正位两边设有配享牌位，称"十二哲"。

崇圣殿
The Chongsheng Hall

碑亭
The Tablet Pavilion

大成门内为中心庙院，院内青砖铺地，古柏参天。中间一条笔直的甬道通向大成殿，甬道两旁浓荫掩映着11座明清纪功碑亭。甬道西南，设有祭奠焚纸用的燎炉和一眼古井。古井当年水深而甘洌，相传当时文人如能饮一杯孔庙古井里的"圣水"，就能笔下生花，文思如泉涌。

清乾隆皇帝赐名为"砚水湖"。大成门外有3座碑，即"明英宗建大学碑"、"清乾隆三十四年(1769年)重修先师庙并颁周彝器谕旨碑"和"清道光九年(1829年)平定回疆告成太学御制文"。

触奸柏
The Chujian Cypress

碑林
The Forest of Steles

刘墉碑
Liu Yong's Stele

碑林
The Forest of Steles

# 历代帝王庙
# The Temple of Successive Emperors

历代帝王庙位于北京市西城区阜成门内大街131号，是明、清时期祭祀三皇五帝和历代帝王、功臣名将的皇家庙宇。1996年被公布为全国重点文物保护单位。

历代帝王庙始建于明嘉靖九年（1530年），清雍正七年（1729年）重修并建碑亭，立御碑。乾隆二十七年（1762年）再次修葺，乾隆二十九年（1764年）完成，同时建碑亭，立御碑，并将景德崇圣殿殿顶的绿琉璃筒瓦易为黄琉璃筒瓦，提高了庙的等级。民国改元，祀典遂废，历代帝王庙改由中华教育促进会及幼稚女子师范学校等单位使用。该庙已于2004年对外开放。

历代帝王庙坐北朝南，占地面积21500平方米，古建筑面积近6000平方米。建筑主体布局分为中、东、西三路，中轴线由南向北依次为琉璃影壁、木牌坊（已拆）、庙门、钟楼、景德门、景德崇圣殿等建筑，中轴线两侧建有配殿。景德崇圣殿是历代帝王庙的主体建筑，大殿东西两侧分别为清雍正、乾隆二帝所建碑亭及御碑。除中路外，东路尚保留有神厨、神库、宰牲亭、井亭等建筑，西路主要为承祭官置斋所配房。

历代帝王庙是我国现存唯一的专门祭祀历代帝王的庙宇。它不仅反映了中华民族悠久的历史，而且也体现了我国统一的多民族国家一脉相承的历史特点。此外，历代帝王庙对于研究古代建筑、封建典章制度均具有较高的文物历史价值。

庙门
The Front Gate

景德门
The Jingde Gate

Located at No.131 Fuchengmennei Street in Xicheng District, the Temple of Successive Emperors was used in the Ming and Qing Dynasties to offer sacrifices to successive emperors, persons who had rendered outstanding service, ancient heroes and the Yellow Emperor. First built in the 9th year of the reign of Emperor Jiajing of the Ming Dynasty (1530 A.D.), it was listed as a national key relic under special preservation in 1996.

Facing south the temple covers an area of 21,500 square meters. The main buildings are divided into the middle, the eastern, and the western axes. On the middle axis the principle structures, from the south to the north, are the glazed screen wall, the Front Gate, the Bell Tower, the Jingde Gate and so on. There are tablet pavilions standing on each side of the main hall.

碑亭
The Tablet Pavilion

景德崇圣殿
The Jingde Chongsheng Hall

大殿面阔九间、重檐庑殿顶，顶覆黄琉璃筒瓦，檐下施以斗拱，木构
架绘以和玺彩画。殿前有一长方形月台，月台东、南、西三面有汉白
玉石护栏，南面三出陛，中为御路，其规格之高，仅次于故宫太和
殿。殿内原有11龛供历代帝王牌位。大殿整体建筑宏大，与东西两庑
组成祭祀之所。

井亭
The Well Pavilion

三出陛石栏台阶
Stone Steps

燎炉
The Sacrificial
Paper Burner

园林

GARDENS

俯鏡清流

# 北京的园林

我国的造园历史相当久远，且园林数量众多，按其隶属关系划分主要有四类：皇家园林、私家园林、景观园林、寺观园林，其中北方的皇家园林与南方的私家园林是中国园林的代表。北京园林是指北京地区的园林，它属于北方园林的一支；北京园林大多是帝王宫苑，具有皇家气派，同时，也吸取了江南园林的造园手法。尽管它们历经沧桑，有过繁荣和昌盛，也遭到了无数次毁坏和洗劫。但是，遗留到今天的这部分，仍不愧为世界园林中的珍品。

## 一、北京园林发展简史

### 1.战国至隋唐时期北京地区的园林

北京地区从西周开始，直至春秋战国时期，属于燕国的领地，《战国策·燕策》载：燕伐齐胜后，"蓟丘之植，植于汶篁。"这当是燕国都邑蓟城的园林，也就是北京地区最早的园林。此外，据记载当时曾在今北京西南一带的古城、涿县、易县等处，营建了黄金台、碣石宫、展台、武阳台等多处台观宫苑。秦代，在北京未及营建林苑。大葆台西汉墓出土文物中，发现塑有各种类型庄园的陶器，庄园内有亭、台、楼、榭等，说明汉代时北京地区的园林很盛行。魏晋十六国北朝时，北京当时称为蓟城，北郊一带"亭台远瞩，为燕之旧地"，其位置即今莲花池、玉渊潭、紫竹院一带，河湖纵横，是一处风景优美的地区。隋代蓟城为北方军事重镇，称幽州。隋炀帝大业三年(607年)改称涿郡，隋炀帝曾命当时著名的建筑艺术家阎毗在桑干河北岸风景区(今大兴区)建造了一座离宫，命名"临朔宫"，为皇帝驻跸之所，内陈大量珍宝。此外，隋代北京寺庙园林也开始发展。例如：隋代的白马寺在今右安门一带，至今仍有经幢；隋僧静琬创建了石经山；隋之姚彬盗马庙在今天坛位置上。唐朝，涿郡改称幽州。贞观十九年(645年)，唐太宗远征辽东，亲自统率主力军往返于幽州城。兵败回师，为稳定军心，在幽州城东墙内修建了一座纪念阵亡将士的悯忠寺(今法源寺)，悯忠寺便是一座典型的寺观园林。唐朝中叶以后，藩镇割据，幽州为北方军事重镇，历

任总管(或都督)的地位和权势都很大，有不少是亲王兼任其职。他们利用权势在幽州城郊营建起豪华的园亭别馆，其中最有名的要算海子园。此外，寺观园林在唐代最为盛行，遗存至今的旧址尚有数处。

### 2.辽金时期北京地区的园林

唐末五代后，出现辽宋南北对峙局面，北京地区一直处于辽王朝的势力范围。938—1122年，契丹族建立的辽朝把幽州改称为南京，又称燕京，其历史地位从原来的北方军事重镇逐渐过渡为全国的政治中心。辽代曾在北京地区建瑶屿行宫，有瑶池殿、临永殿。

金朝统治者定都燕京，改燕京为中都，在营建宫殿的同时，金开始在中都城内建造苑囿。中都城内苑囿以东、西、南、北四苑为主，还有其他小型园林。其中又以皇城内之西苑最为主要，由于地近宫城，是金帝及皇室经常游玩的场所。它引西湖(今莲花池)水，苑内的水泊，统称为太液池，池中之岛称为琼华岛。皇城内西苑亦称西园，它还包括同乐园和宫内琼林苑等建筑景物。此外在中都城内还建有东苑(东园、东明园)、南苑(南园、熙春园)和北苑等皇家园林。

在改建中都城之后，统治者又着手营造离宫禁苑。其中规模最大的是万宁宫，其址在中都城东北郊，即今北海公园位置。除万宁宫外，还在都城远近郊建造了几十处离宫、苑囿。金章宗的八大水院，现在能考出名称和地址的只有三处，即金水院(颐和园内)、清水院(大觉寺)、香水院(妙高峰七王坟)，其余五处尚待考证。像玉泉山芙蓉殿金章宗行宫、香山金代行宫和梦感泉、樱桃沟金章宗观花台、潭柘寺附近的金章宗弹雀处、玉渊潭钓鱼台等都是金朝皇帝，特别是金章宗的常幸处所。另有金章宗明昌年间命名的"燕京八景"，也是北京地区历史上有名的景观，最早见于金朝《明朝遗事》中，即：居庸叠翠、玉泉垂虹、太液秋风、琼岛春阴、蓟门飞雨、卢沟晓月、西山积雪和金台夕照。

由此可见，金代是北京园林史奠定早期基础的时期，

它的园林艺术特点，大体是"北宋山水宫苑"的延续。

### 3.元朝北京地区的园林

元世祖至元四年(1267年)，定都燕京，弃中都旧城，在中都城东北郊另建新城，命名大都，城的南北轴线偏前方为宫殿，宫殿西侧即为元代集中精力营建的园林。

还在大都城营建之前，中统三年(1162年)，元世祖忽必烈就将金代在中都城外的离宫加以整修，作为临时驻跸的处所。这座离宫就是金中都东北郊的大宁宫，后改为万宁宫，由湖泊、岛屿和宫殿组成。湖中小岛是琼华岛，岛上建广寒殿，山腰有仁智殿，琼华岛对面水中的团城上有仪天殿。宫城建成以后，琼华岛称为万岁山，水面改称为太液池。山池总体仍然体现了秦汉以来仿海上神山的传统，奠定了明清三海的基础。元代万岁山上的遗物"渎山大玉海"玉瓮，原放置在广寒殿内，专供忽必烈贮酒饮宴，后来流落于民间，清代被乾隆帝访得，移置团城的玉瓮亭内，供人欣赏。

元代皇家园林虽有发展，但规模不大，这时期在城近郊泉池地带分布着一些权贵和士大夫们的私园，他们的造园活动超过了以往的各个朝代，其中较知名的有漱芳亭、万春园、蓬莱观、松鹤堂、万柳堂等等几十处。据记载，漱芳亭为最早引进南方梅花到北京种植的地方。

### 4.明清时期北京地区的园林

明清两代，尤其是明代的嘉庆至清乾隆朝，由于商业繁荣兴盛，这个时期是北京乃至全国古典园林的鼎盛时期，不仅园林数量众多，在造园艺术上也达到了极高水平。

明代皇家园林建设重点在大内御苑，除少数(如御花园)建置在紫禁城的内廷，多数都建置在紫禁城以外、皇城以内的地段，以便皇帝经常游幸。这一时期的皇家园林，仍以万岁山、太液池为主。当时太液池向南展拓，成为北海、中海、南海三海一贯的水域，总称西苑，与新建的紫禁城南北向长短相当，在紫禁城西部，形成了一道水面屏障，在三海沿岸和池中岛上增建殿宇，总称西苑，与紫禁城之间只有一条长街隔开，构成宫苑相连的宏大布局。明代的皇家园林中还有东苑和南苑，其中南苑位于城南郊，古称南海子，麋鹿黄羊出没其中，风景优美，是著名的燕京十景之一——"南囿秋风"。

由于明成祖迁都北京后，随着政治中心的北移，北京逐渐成为北方的佛教中心，寺观建筑逐年有所增加，加大了寺观园林的发展。北京的西北郊，大量兴建佛寺，一般都建有园林，例如香山寺、碧云寺。

此外，明代城区和近郊修建了不少官僚士大夫们的府邸宅院，较著名的有：定国公园、英国公新园、李皇亲之钓鱼台、惠安伯园、勺园、湛园、梁梦龙园、李园(清华园)等等。

清王朝定都北京后，对皇家园林的兴建一直没有间断，自康熙、雍正和乾隆祖孙三代，前后130年，相继在西郊海淀以北东西10公里内建成了规模宏大的"三山五园"皇家园林区，即作为皇帝长期居住、进行政治活动的离宫御苑，如圆明园、畅春园；作为皇帝短期驻跸游玩的行宫御苑，如香山静宜园、玉泉山静明园、万寿山清漪园(颐和园)。圆明园面积最广，在五园中最为有名，园内楼台亭榭，步移景换，全园有108个景点，是一处规模宏大的皇家园林。它不仅有中国南北方的山水景色，还引入了把西方造景的观念，直接把西洋建筑和景物布置在园子里。在"三山五园"区内，还相继修建了皇亲国戚、王公大臣的赐园及宅园，最有名的是漱春园，其旧址在北京大学未名湖畔，附近相互毗连的有镜春园、鸣鹤园、朗润园、蔚秀园、承泽园等。

与此同时，紫禁城内，重葺御花园、建福宫西御花园、宁寿宫花园和慈宁宫花园；在紫禁城北景山五峰之上，增建五亭；在西苑"三海"之中，增建、修葺甚多，如北海阐福寺、万佛楼、中海紫光阁、南海宝月楼等等。

北京曾是几代皇朝都城，皇家的子孙长大后都要分封一个王府。这些王府除了建造厅堂楼阁外，还要修建花园，而且无论距离河道多远，也都千方百计把河水引入园内。明代以前的王府大都湮没或改建，清代所封王爷近百，他们的王府花园也有几十座，今天保留下来的几十座私家园林大都是清代的王府花园或皇家近亲的宅舍别业。最有名的当是恭王府及花园。

咸丰十年(1860年)英法联军侵逼北京，焚毁了建造百余年的圆明园，清漪园、静明园也遭劫掠，致使清代园林精华全部损失，也是我国园林史上的重大损失。至此，中国封建社会的最后繁荣阶段已经结束，皇室已没有财力营建新的园林，北京的园林建设开始衰落。

# 二、北京园林的特色

北京的园林属于典型的北方园林，它基本上可以代表中国园林的风格与特色，既不拘泥于风景的布置，也不重点修建过多的楼、阁、亭、榭，而是以其自然活泼的配置方式为建园的基础。

## 1.自然特点

北京地处广阔的华北平原，三面围山，虽然地势平坦开阔，但可利用的河川、湖泊比南方少，湖泊、水源和水量的限制是北京园林的制约因素。北京最美的园林在海淀，海淀之美，美在水。海淀园林的开发和建设充分利用了丰富的水源条件，无论是设计小型私家园林还是规划大型皇家园林，对水的利用和开发都达到了极致。颐和园利用了天然山水之盛景，巧夺天工地布置了各景区的造型。圆明园没有大的自然水面，就挖湖堆山，植树披绿，以匀称合理的布局充分体现了人的意志。静明园和静宜园则有意识地把地下泉水提高到显著位置，让青山与碧水和谐统一。除了皇家园林占有湖泊之外，一些王亲贵族的私家园林只能得到皇家的残羹冷炙，水面不大，但也做足了水文章，而有些只能是平地造旱园，即使园中有水，也是水面很小。相对于全园面积，北京园林的水面比率是很少的。

北京园林还具有崇山性，表现在园林的堆山上，以高、壮为美，山体面积较大。清代皇家及贵族有权力、财力以及人力营造如此雄伟的山，如北海的琼华岛、御花园的堆秀山、景山公园的景山、恭王府花园的假山等等。

在植物方面，北京园林内常绿阔叶植物少，多以落叶植物为主。北京的皇家园林、寺院、陵园建筑为了反映帝王的至高无上的权利、等级分明的特点，常选择姿态苍劲、意境深远的中国传统树种，如白皮松、油松、圆柏、国槐、银杏、丁香、海棠、牡丹、芍药、荷花等。

## 2.布局特点

北京园林的布局普遍遵循前朝后寝、轴线对称、一池三山、仿景缩景、障景漏景等造园原则，体现出儒道佛三家对园林的渗透。

前朝后寝在园林中主要表现为两方面：一是园与宅关系上，园林居于宫殿、住宅的后部或侧位，如故宫的御花园、景山、宁寿宫花园(乾隆花园)、慈宁宫花园都是在后部；二是游览与居住关系上，居住在前，游览在后。这些布局直接与儒家"先天下之忧而忧，后天下之乐而乐"的观点一致。

轴线对称是明、清两代遗留下来的园林最明显的特点。园内建筑物体高大，在全园布局上分前殿、中殿和后殿，采取中轴对称的形式。皇家、私家、寺观皆如此，只不过在轴线和对称的程度上有所差别而已。园林的轴线与宫殿和住宅的轴线一致，成为宅区轴线的延伸。

一池三山是中国古代宫苑建筑中常见的规划形式，通常表现为在一片水域中布置三座岛屿，它源于齐国的东海神话，东海里有蓬莱、方丈、瀛海三座仙山，山上长满了长生不老药，住着长寿快乐的神仙。封建帝王都梦想万寿无疆与长久统治，自从汉武帝在长安城修建了象征性的"瑶池三仙山"开始，"一池三山"就成为历代皇家园林的传统格局。北京园林的一池三山在金、元、明、清四朝表现在西苑三海景点创作上，一海一山，南北一线，成一池三山之制。颐和园有大小三山，大三山是：南湖岛、藻鉴堂、治镜阁，小三山是：凤凰墩、知春亭、小西冷。此外，在私家园林和寺观园林中也多有在园中凿池，并在池中布置一岛、二岛或三岛以象征仙山的事例。

仿景缩景。仿景常指原比例摹仿，缩景常指变比例摹仿，清代康、乾二帝多次南巡，深受南方清雅秀丽的山水园林影响，并在北京皇家园林里加以效仿，因此皇家园林中也出现了顺应自然的园林建筑和模拟江南名园的园中之园，于是异地仿景和缩景达到了高潮。清乾隆皇帝对以江南园林为蓝本，仿建于御园之内的景致，是这样评价的："略师其意，就其自然之势，不舍己之所长"，即重在求神似而不拘泥于形式，是运用北方刚健之笔抒写江南柔媚之情的一种更为难能可贵的艺术创造。圆明园内的"做石临流"一景，通过三面人工假山，引水成瀑潴而为小溪的布局，来凝缩、移植和摹写浙江绍兴兰亭著名的崇山峻岭的构思；以杭州西湖"玉泉观鱼"的鱼泉相戏，悠然自得的景色，再现在圆明园内的"坦坦荡荡"之中。当然，这样的江南园林在北京的再现，决不是简单摹仿，而是结合北京的自然条件，使用北方的材料，适应北方的鉴赏习惯的一种艺术再创造。另外，统治者不惜重金调集全国各地的有用之材修建苑囿，体现了浑厚宏大的皇家气派，这是南方别致玲珑的园林无法相比的。修建颐和园时，将这一优势发挥到淋漓尽致的程度。由于北京皇家园林有私家园林难以兼备的大环境和人力、物力的基础，并在"移天缩地在君怀"的思想指导下，把江南风景名胜园林加以精炼和再创造，常常建筑不多却颇有神韵，并兼具南北之长。

障景和漏景。障景是园林造景手法之一，即在园林中设置屏障似的景观或景物。北京园林的障景表现为严密性，从围墙的障景上看，大凡墙隔较少漏窗，即使有漏窗，也是较为厚重的花式或直接玻璃屏蔽。漏景是通过漏明墙、漏窗摄取景观的造景手法，按道理说是缩小了园林空间，但实际上正是这种看似缩小的取景手法，反倒是扩大了园林的空间，丰富了园林景观。

从设计要素上看，北京园林的小品、装饰、建筑、堆石都有很强的特点。北京园林有盆景奇石、石狮等，装饰上彩画有和玺彩画、旋子彩画、苏式彩画等。在保持北方建筑传统风格的基础上，还借鉴了江南园林的造园手法，大量使用游廊、水廊、爬山廊、拱桥、亭桥、平桥、舫、榭、粉墙、漏窗等江南常见的园林建筑形式以及堆叠假山的技法。

总之，北京园林在风格上，将江南造园手法融于北方特色之中，继承和发扬了造园艺术，又有创新，北京园林的全部历史反映了这些特点的成长过程。

# Gardens in Beijing

The garden making in China has a long history, and established a flagship, enjoying a higher reputation. The royal gardens in the northern China and the private gardens in the southern China are representative of Chinese gardens. Most of the gardens in Beijing are the palaces of the emperors with royal style, and also absorbed the garden making techniques used in the gardens in the south of Yangtze River. Although these gardens underwent great changes, flourish and prosperity, as well as numerous times of damages and loots. Nevertheless, the gardens left to present day are still the precious treasures in the gardens in the world.

## 1. The Brief History of the Development of Beijing's Gardens

### (1) The Beijing gardens built from the Warring States Period to the Sui and Tang Dynasties

From the beginning of the Western Zhou Dynasty to the Spring and Autumn and Warring States Periods, the Beijing area belonged to the territory of the Yan State. The gardens built in Beijing at that time belonged to the Ji City, capital of the Yan State, and was the earliest form of gardens in the Beijing area. As seen from the unearthed cultural relics from the Dabaotai tombs of the Western Han Dynasty, all types of potteries shaped like all types of manors were found there. In these manors were pavilions, terraces, buildings, houses etc., indicating gardens were very prevalent in Beijing during the Han Dynasty. In the Wei, Jin, Northern Dynasty, Beijing was called the Ji City at that time, and was located in the present-day area of Lianhuachi, Yuyuantan, and Zizhuyuan, presenting a beautiful scenery with the rivers and lakes in a crisscross pattern. The Emperor Suiyang of the Sui Dynasty ordered Yan Pi, a famous architectural artist, to build a temporary palace, called "Lin Shuo Palace" in the scenic spot in the northern bank of the Sangqian River. The palace was a place for the emperor to be stationed, and kept a large amount of treasures. In addition, the temple gardens began to evolve in Beijing during the Sui Dynasty. The Minzhong Temple, a typical temple garden, was built inside the eastern wall of Youzhou City, for memory of deceased warriors. After the middle period of the Tang Dynasty, a luxurious garden and annexes were erected in the outskirt in the Youzhou City, of which the most famous one was the "Haizi Garden". The temple gardens were most popular in the Tang Dynasty, and there are still several relics in existence nowadays.

### (2) The gardens in Beijing during the Liao and Jin Dynasties

The Yaoyu Temporary Palace was built in the Beijing area in the Liao Dynasty, with the Yaochi Hall and the Linyong Hall constructed in the palace. The rulers of the Jin Dynasty located its capital in Yanjing whose name was changed as Zhongdu. While building the palace, four major gardens, in the east, west, south, and north respectively, as well as other small-scale gardens, were established in the Zhongdu City. Xiyuan Garden inside the imperial city was the primary one.

After Zhongdu was rebuilt, the rulers set out constructing the forbidden gardens of the temporary palace, of which, the Wanning Palace with largest effect was situated in the northeastern suburb of Zhongdu, i.e. the present-day Beihai Park. In addition to the Wanning Palace, dozens of temporary palaces and gardens were established in the faraway or nearby suburbs in the capital city. *The Eight Sceneries in Yanjing* which were formed during the reign of Emperor Zhangzong of the Jin Dynasty, was also the famous scenic spots in Beijing. Thus, it can be seen that the Jin Dynasty was a period which laid an early foundation for the evolvement of gardens in Beijing. The garden artistic characteristics were an extension of the palaces and gardens with hills and water built in the Northern Song Dynasty.

### (3) The Gardens in Beijing during the Yuan Dynasty

Before Dadu was built, kublai, the first emperor of the Yuan Dynasty, had repaired the Daning Palace in northeastern Jin Zhongdu. The palace was later renamed Wanning Palace, which consists of lake, island and palace. After the palace was built up, Qionghua Island was renamed Longevity Hill and the lake was renamed Taiye Pool. Imperial gardens during the Yuan Dynasty made a progress but their scale was not large. In the near suburbs of the city were also distributed some private gardens of dignitaries and scholar-bureaucrats. Scores of these gardens were famous, such as Shufang Pavilion, Wanchun Garden, Hulutao, Penglaiguan, Songhe Hall, Wanliu Hall etc.

### (4) The Gardens in Beijing during the Ming and Qing Dynasties

Construction of royal gardens during the Ming Dynasty focused on Danei Imperial Garden. Except few architecture (like imperial garden) built in the Inner Court of the Forbidden City, most architecture were built outside the Forbidden City and inside the imperial city, so that the emperor could visit them from time to time. Imperial gardens during this period were still represented by Longevity Hill and Taiye Pool, which were improved further.

The largest sacrificial garden during the Ming Dynasty was the Tombs of 13 Ming Emperors, the Ming Tombs. Garden of the Ming Tombs extended more 40 kilometers along the natural topography of the hill and the architecture were well laid out in a symmetric way. Hence, the whole garden looked just like a beautiful picture.

During the Ming Dynasty, a lot of residences of officers and scholar-bureaucrats were built in the urban and near suburb areas. Some of them were famous, such as Dingguo Park, Yingguogong New Garden, Fishing Platform of Emperor's Relative Li, Huianbo Garden, Shao Garden, Zhan Garden, Liangmenglong Garden, Li Garden

(Tsinghua Garden) etc.

After the Qing Dynasty settled its capital in Beijing, three emperors: Kangxi, Yongzhen and Qianlong spent a total of 130 years building a large imperial garden complex, which consists of "Three Hills and Five Gardens" and runs 10 kilometers from the east to the west in the western suburb to the north of Haidian District. Among the five gardens, Yuanmingyuan was the largest and best known among the five gardens. In this super-large imperial garden gathered 108 landscapes including pavilions, terraces and towers, so visitors can feast their eyes on different sceneries as they go around. By combining Western style of scenic design into the landscapes typical of the north and the south of China, Yuanmingyuan had western architecture and landscapes directly arranged together. In the "Three Hills and Five Gardens" were also built some granted yards and residences of emperor's relatives and senior officials. The best known of them was Shuchun Garden, the site of which is located in the bank of Weiming Lake, Peking University, and adjacent to it were Jingchun Garden, Minghe Garden, Langrun Garden, Weixie Garden, Chengze Garden and so on.

At the same time, in the Forbidden City, the Imperial Garden, the Imperial Garden to the West of Jianfu Palace, Ningshou Palace Garden and Cining Palace Garden were renovated; five pavilions were newly built on the five peaks of Jingshan Hill north to the Forbidden City; in the "Three Seas" of the Xiyuan Garden, a lot of buildings were newly built and renovated, such as Shanfu Temple and Buddha Mansion in the North Sea, the Tower of Vermilion Light in the Central Sea, the Precious Moon Tower in the South Sea and so on.

## 2. The Characteristics of Beijing Gardens

Beijing garden belongs to typical northern garden. Basically it can represent the style and characteristic of

Chinese gardens, which are constructed on the basis of natural and lively arrangement rather than rigidly adhere to the arrangement of the landscape or emphatically establish abundant pavilions, terraces, and towers.

### (1) The natural characteristics

Beijing locates in the capacious North China Plain and is surrounded by mountains in three directions. Although it is level and open in topography, the available rivers and lakes are less than those of the south. Thus, lakes, water resources and the amount of water become the limitation of Beijing garden. The most beautiful garden in Beijing is in Haidian District and the beauty of Haidian is in its water. The exploration and the construction of gardens in Haidian District made full use of its abundant water resources. No matter it's the design of a small-sized private garden or the programming of a large-scale imperial garden, the exploitation and development of water reached its perfection. The worshiping for hills in Beijing garden displays in the piling up of rockeries. The majesty of garden hills lies in its highness and greatness. The hills are relatively big in volume and high in highness. It's true that the royal and the noble in the Qing Dynasty had enough power, wealth and human resource to build hills of such majesty.

In the perspective of plants, there are less evergreen broad-leaved plants with deciduous plants taking the biggest proportion. The plants in imperial gardens, monasteries, and cemeteries in Beijing are normally those traditional tree species of great magnificence in order to reflect the supreme power of the emperors and the strictness of hierarchy. These trees include lacebark pine, Chinese pine, cypress, Japanese pagoda tree, Chinese Horse-chestnut, ginkgo, clove, Chinese flowering crab-apple, peony, Chinese herbaceous peony and water lily.

### (2) The characteristics of the layout of Beijing gardens

The layout of Beijing gardens normally follows the principle of putting gardens behind living places, axial symmetry, one pond surrounded by three hills, imitating a scene and zooming it and covering and leaving out a scene, which reflect the influences of Confucianism, Taoism and Buddhism.

The principle of putting gardens behind living places means the backward layout of gardens. This mainly embodies in two aspects; first, in the respect of the relationship between garden and residence, garden should be in the back or beside palace or residence; secondly, in the respect of the relationship between sightseeing and living, living should be more important than sightseeing. These principles of layout are in direct connection with the Confucian view of "Be the first to feel concern about state affairs and the last to enjoy oneself!"

The principle of axial symmetry is the most obvious characteristic of the gardens left behind by the Ming and Qing Dynasties. The axes of gardens are in accord with those of palaces and residences and become the extension of the axes of the residences. On the central axis line the most important gate, hall, palace, corridor and pond. Among all the axes, the most magnificent one is that of the Summer Palace. Cining Palace (the Hall of Benevolent Peace) Garden, Le Family Garden, Prince Gong's Residence Garden and Jingshan Hill all have one axis. However, some gardens adopt many axes, for instance, there are the middle, the eastern and the western axes in the Imperial Garden and there are the eastern and the western axes in the Jianfu Palace Garden.

In some of the gardens, there are not only axes but also symmetry. The symmetry of the Summer Palace rests mainly in the formation of Longevity Hill and the symmetry of Yuanmingyuan Garden manifests itself in the ten sceneries of West Lake in Fuhai (the Sea of Auspiciousness). The four gardens in the Imperial Palace and Jingshan Hill are all strictly in symmetry while sceneries in some other gardens like Prince Gong's Residence Garden, Keyuan Garden and the Garden of Le Family are in symmetry in different levels. This symmetrical idea is in direct accord with the concept of relative supplement of Yin and Yang in Taoism and the relative complement of civil officials and military officials in Confucianism.

The term one pool surrounded by three hills comes from the East Sea myth and realizes in the concept of the longing for longevity in building imperial gardens. This principle in Beijing gardens mainly displays in imperial gardens, for instance, the four Dynasties of Jin, Yuan, Ming and Qing created one sea and one hill with one line traversing from the south to the north in the Three Seas scenery in Xiyuan, which formed the system of one pool surrounded by three hills.

Imitating a scene often refers to copying without change in proportion while zooming a scene means copying with changes in proportion. In the Qing Dynasty, the Emperors Kangxi and Qianlong had patrolled south for several times and deeply influenced by the fresh, elegant, delicate and beautiful landscapes and gardens. They spent time copying them in Beijing imperial gardens. In this way, natural garden architecture and famous southern gardens appeared in rigid and majestic imperial gardens. Imitating a scene and zooming it in a different place reached their summits.

The principle of covering a scene in Beijing gardens embodies itself in its strictness. Seeing from the walls, few walls have windows, if they have, they are of massive style or direct glass screens. The covers before entrances are mostly walls decorated with dragons. Since most of the extant gardens are imperial gardens, the Nine Dragons Wall in the North Sea becomes the representative of Chinese screen walls.

# 北海及团城
# Beihai and Round City

　　北海及团城位于故宫西北、景山西侧。北海既是辽、金、元、明、清五代帝王的宫苑，又是我国保存至今历史最悠久、规模宏伟、布置精美的古代园林杰作之一。1961年被公布为全国重点文物保护单位。

　　北海始建于辽代，辽代时仿照蓬莱仙境建瑶玙行宫。

金代时在此建大宁宫，后改万宁宫，又改岛名为琼华岛，在山顶建广寒殿。元代三次扩建琼华岛，并以岛为中心修建了帝都和宫苑。明代北海、中海、南海合称三海，亦称太液池，并列为禁苑，后在太液池北岸修筑五龙亭。清顺治八年（1651年）在广寒殿旧址建造白塔，并将岛南部的宫

殿改建为永安寺。乾隆时除在琼华岛四面建亭榭楼台外，又在北岸修建了先蚕坛、阐福寺、西天梵境、万佛楼、小西天、澂观堂、镜清斋（今静心斋）等，在东岸修建濠濮涧、画舫斋等，具备了今天北海的规模。民国初年，中海和南海合并为一园，称中南海，北海另辟一园。

北海总面积约为70公顷，水面占2/3，全园布局以琼岛和岛上的白塔山为中心。岛上的建筑依山而建，按其布局大致可分为以白塔为主体的东、南、西、北四个部分。琼岛的顶部矗立着35.9米高的藏式白塔。山的南坡以永安寺为主体，有法轮殿、正觉殿、普安殿、钟鼓楼等建筑。永安寺西的悦心殿是皇帝引见群臣和处理朝政的地方，其后的庆霄楼是帝、后冬日观冰嬉戏之处。山的西坡有琳光殿和阅古楼等建筑，阅古楼是皇帝私人藏书的地方，楼平面呈半月形，墙壁遍嵌《三希堂》法帖刻石495方，保留了魏晋以来书法家的墨迹。山的北坡有仿镇江金山寺而建的漪澜堂和道宁斋、碧照楼、远帆阁、延楼等，与太液池北岸五龙亭、西天梵境隔水相望，构成太液池畔交相辉映的两组重要建筑群。山的东坡自山门、石桥、牌坊至智珠殿、见春亭，构成燕京八景之一的"琼岛春阴"，至今还保存有清乾隆皇帝题诗碑。

太液池东、北沿岸分布有画舫斋、濠濮涧等建筑。

静心斋是北海北岸的主体建筑，规模不大，院落布局紧凑，曾是清朝太子读书的地方，乾隆皇帝常在这里抚琴吟诵，故又称乾隆小花园。此外，北海北岸还有小西天、大西天、九龙壁、五龙亭、阐福寺等建筑。

团城在北海公园南门西侧，位于北海与中海金鳌玉蛛桥东桥头北侧。团城既是北海的一部分，又是一个具有独特风格的独立的小园林。团城原是太液池中的一个小岛，金大定三年至十九年（1163—1179年），金世宗完颜雍于岛上开始建造宫殿，成为御苑的一部分。明清两代增建了承光殿、玉瓮亭、古籍堂、敬跻堂、余清斋、镜澜亭等建筑，并在岛屿周围加筑城墙。团城高5米，面积约4500平方米。承光殿是团城的主体建筑，殿内雕龙佛龛中供奉着一尊白玉释迦牟尼坐佛。殿前正中有琉璃方亭，亭内陈设元至元二年（1265年）所雕玉瓮，名"渎山大玉海"，元世祖忽必烈曾用玉瓮盛酒，宴赏功臣。明代玉瓮流落于民间，清代被乾隆帝访得，移置团城，并建亭保护，并亲笔撰写《玉瓮歌》、《玉瓮诗》铭刻在玉瓮上。

牌楼
The Archway

To the northwest of the Forbidden City and the west of Jingshan Hill lie the Beihai and Round City, which are located at No.1 Wenjing Street in Xicheng District of Beijing. The Beihai was the pleasure palace of Liao, Jin, Yuan, Ming and Qing emperors and is one of the earliest imperial palaces extant in China. It was listed as a national key relic under special preservation in 1961.

The Beihai was first constructed in the Liao Dynasty. The Liao Dynasty built the Yaoyu Palace by imitating the Penglai Fairyland. The lake was enlarged in the Jin Dynasty, using rocks for piling on the hill and building the Guanghan Palace. In the Yuan Dynasty, the Qionghua Islet was enlarged three times, and the imperial residence and palaces were built with the islet in the center. In the Ming Dynasty, Beihai (North Sea), Zhonghai (Central Sea) and Nanhai (South Sea) were collectively named as Three Seas or the Taiye Lake, which was the forbidden garden. Then, the Five-Dragon Pavilions were built on the northern bank of the Taiye Lake. In the 8th year of the reign of Emperor Shunzhi of the Qing Dynasty (1651 A.D.), the White Dagoba was first built on the site of the Guanghan Palace and palaces on the south of the islet were rebuilt to the Yong'an Temple. The reign of Emperor Qianlong built the Silkworm Altar, the Hall of Heavenly Kings, the Wanfo Tower and the Quieting Heart Chamber on the northern bank and built the Huafang Chamber and the Shemei Chamber on the eastern bank.

The Beihai has an area of about 70 hectares, more than half of which is water. The Qiong Islet and the White Dagoba on it is the center of the whole park. The architectures on the islet were constructed in accordance with the shape of the White Dagoba Hill, matching with each other horizontally and vertically. With the White Dagoba as the landmark, the layout of the islet can be divided into east, south, west and north parts. On the top of the Qiong Islet stands the Tibetan-style White Dagoba which is 35.9 meters high. On the south of the hill stands a main building, the Yong'an Temple. The Linguang

(Sunshine) Hall and the Yuegu (Reading the Classics) Building sit on the west of the hill. On the hill slope, there are the Mujian Chamber, the Yanyun Jintai Pavilion and other ancient buildings, antique and tranquil. On the bank of the Taiye Lake there are two important building compounds that stand opposite to each other. One compound lies on the north of hill, including the Yilan Hall, the Daoning Chamber, the Bizhao Building, the Yuanfan Pavilion, the Yan Building, etc. The other lies on

琼岛远景
The Qiong Islet Viewed from Far

the north of the Taiye Lake, including the Wulong Pavilion and the Xitian Fanjing Pavilion. On the east of the hill there are flourishing woods, rocks, deep caves, stone bridge, memorial archway, the Zhizhu (Intellectual Pearl) Hall and the Jianchu Pavilion, forming the Qiong Dao Chun Yin (Jade Islet's Spring Shadow), one of the eight beautiful sceneries in ancient Beijing. The tablet, with Qing Emperor Qianlong's inscription on it, is well preserved till today.

On the east and north of the Taiye Lake, pavilions and halls are hidden in green woods and waves. The Huafang Chamber is exquisite and graceful, and the Long Corridor winds its way deeply. That place can rival the wonderful gardens in Southeast China. To the north of the Huafang Chamber is the Haopu Mountain Stream, which faces the lake on its three sides. The stream is surrounded by artificial hills and zigzagging stone bridges, presenting a special view. The Quieting Heart Chamber is the main

永安寺
The Yong'an Temple

永安寺有法轮殿、正觉殿、普安殿、钟鼓楼等建筑，这些建筑均为歇山顶，并覆盖有黄、绿、紫等各色的琉璃瓦，白山顶俯瞰，色彩斑斓，蔚为壮观。永安寺西的悦心殿是皇帝引见群臣和处理朝政的地方，其后的庆霄楼是帝、后冬日观冰嬉戏之处。

白塔
The White Dagoba

善因殿
The Shanyin Hall

building on the northern shore of the lake, covering an area of one hectare, which was the place where the princes studied. Emperor Qianlong used to play musical instruments and read books there, so the Quieting Heart Chamber is also called Qianlong's Miniature Garden.

The Round City is located in the west of the south gate of the Beihai Park. It lies to the north of the east end of the Golden Tortoise Jade Rainbow Bridge, which is between the Beihai and Zhonghai. The Round City is not only a part of the Beihai Park but also an independent garden with its unique style. Together with Beihai, Zhonghai and Nanhai, it forms the most beautiful scenic area in Beijing City.

Originally, the Round City was an islet in the Taiye Lake. From the 3rd year to the 19th year of the Dading reign of the Jin Dynasty (1163-1179 A.D.), Emperor Shizong started to build palaces on the islet. The Round City, with the Qionghua Islet facing each other at a distance, is a part of the imperial palace. More buildings were constructed in the Ming and Qing Dynasties such as the Chengguang Hall and the Yuweng Pavilion. Besides, the walls and crenels were built around the islet, forming the embryonic shape of the Round City. The Round City, surrounded by 5-meter-high walls, has an area of 4,500 square meters. The Chengguang Hall stands in the middle of the city and is the main building inside.

琼岛鸟瞰
The Bird View of the Qiong Islet

悦心殿
The Yuexin Hall

北海延楼长廊外景
Exterior of the Long Corridor

"琼岛春阴"碑
The Tablet with the Inscription
"Qiong Islet's Spring Shadow"

铜仙承露盘
Bronze Figure

画舫斋
The Huafang Chamber

濠濮涧
The Haopu Mountain Stream

唐槐
Tanghuai

静心斋内景
The Quieting Heart Chamber

全园以山石池水为中心，以斋亭楼轩点缀其间，飞瀑流水，往还萦
回，景致变幻无穷。

静心斋内景
The Quieting Heart Chamber

静心斋内景
The Quieting Heart Chamber

抱素书屋
The Chamber for Reading the
Classics

西天梵境山门琉璃装饰
The Glazed Decoration on the
Front Gate of the Small Sukhavati
Garden

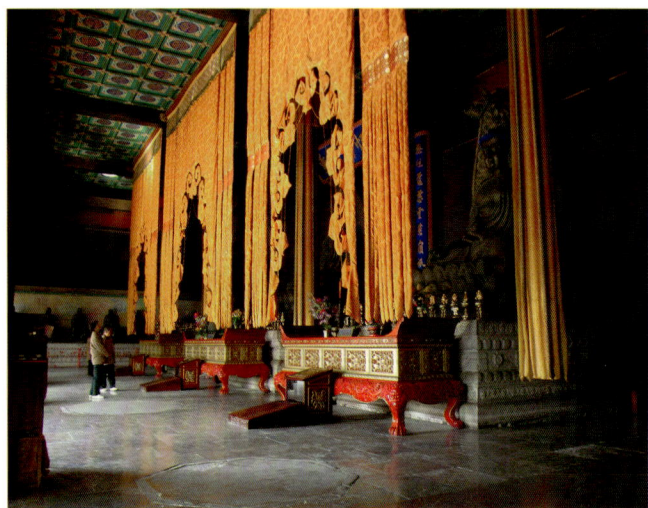

大慈真如宝殿内景
Interior of the Hall of Great Mercy
and Eternal Truth

大慈真如宝殿
The Hall of Great Mercy and
Eternal Truth

九龙壁
Nine-Dragon Screen

九龙壁侧面
The Side of Nine-Dragon Screen

九龙壁(局部)
Section of Nine-Dragon Screen

极乐世界
Miniature Western Heaven

从琉璃牌坊望北海白塔
Looking at the White Dagoba from
the Glazed Archway

五龙亭（局部）
Section of the Five Dragon Pavilion

五龙亭鸟瞰
The Bird View of the Five Dragon Pavilions

北海团城
The Round City of Beihai

团城承光殿
The Hall to Receive the Light

承光殿平面呈"亞"字形，黄琉璃筒瓦，重檐歇
山顶，四面各推出单檐卷棚顶抱厦，色彩绚
丽，装饰十分豪华。
殿内木刻雕龙佛龛中供奉着一尊白玉释迦牟尼
坐佛，为一方整玉雕琢而成。

团城承光殿玉佛
Jade Buddha in the Hall to
Receive the Light

"遮荫侯"古树
The Pine Named the Sunshade
Marquess

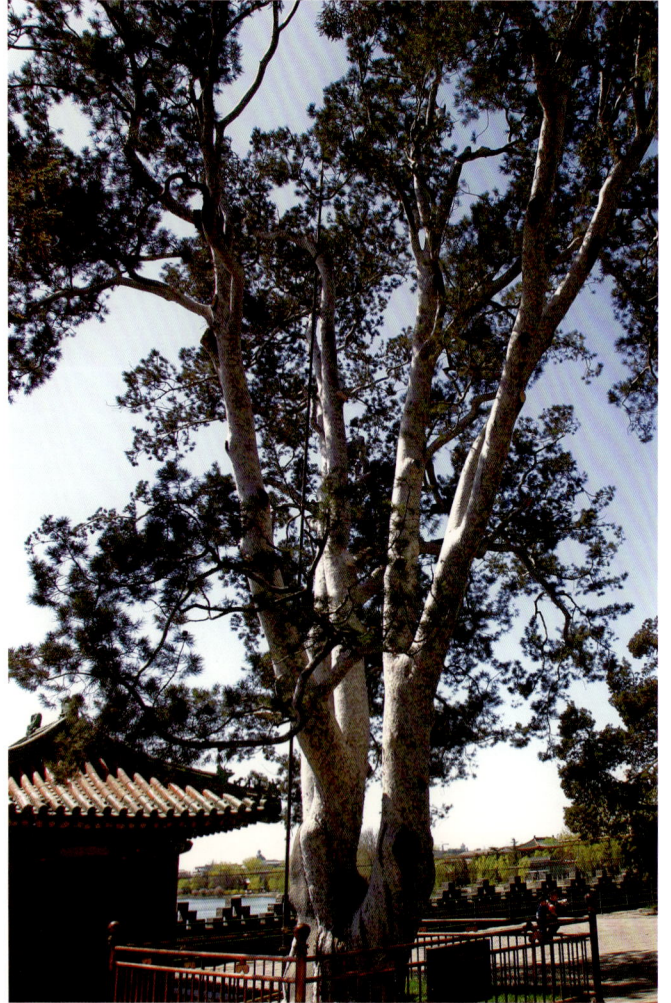

"白袍将军"古树
The Lacebark Pine Named the
General in White Robe

白塔晨辉
The White Dagoba in
the Morning

# 中南海
# Central and South Seas

　　中南海位于北京市西城区西长安街北侧，是中海和南海的统称。中南海是北京皇家建造的禁苑，是一座具有重大历史、艺术价值的皇家园林式建筑群。2006年被公布为全国重点文物保护单位。

　　南海、中海、北海原为一连片水域，始建于辽，金代在此建万宁宫，元代以此为中心营建皇宫大内。明代宫城东移，此处作为宫城以西的禁苑，称西苑，并展拓南海水域。明后期起，始按地域分称南海、中海和北海，又称为前三海(与什刹海等后三海相对而言)，其间，南海与中海以蜈蚣桥为界，中海与北海以金鳌玉蝀桥为界。清沿明制。乾隆年间，大兴土木达30年，形成今天所见园林格局。清后期又进行过全面维修和扩展，形成了在京城中水面最大、建筑最完整、规划最好、规模最大的皇家园囿。

　　中南海总占地面积约为1500亩，其中水面约为700余亩。现存建筑绝大部分为清代遗构。南海主要建筑有宝月楼、瀛台；中海主要景物有紫光阁、蕉园和孤立水中的水云榭。建筑完全按行宫的建筑形式建造，既有宫殿式建筑的庄严肃穆，又有园林式建筑的灵活多变，各景区建筑群相对独立，但又彼此遥相呼应。

迎薰亭(瀛台)
The Sea Terrace Island

瀛台是位于南海之中的半岛，又名南台，东、南、西三面临水，台上为一组殿阁亭台、假山廊榭所组成的水岛景区。瀛台之名取自传说中的东海仙岛瀛洲，寓意人间仙境，是清帝后游幸避暑胜地，也是康熙、乾隆等皇帝居园理朝听政之所。瀛台岛在顺治、康熙时都曾大规模地修建。岛上的建筑物按轴线对称布局，主要建筑都在轴线上，自北至南有勤政殿、翔鸾阁、涵元门、涵元殿、香扆殿、迎薰亭等。与东西朝向的祥辉楼、景星殿、庆云殿等共同组成三重封闭的庭院。沿瀛台岛又点缀了许多赏游的建筑：东面有补桐书屋、随安室、镜光亭、倚丹轩，以及建于水中的牣鱼亭；西面有长春书屋、八音克谐亭、怀抱爽亭等。

Located on the north of Xichang'an Street in Xicheng District, the Central and South Seas refer to two lakes. It was upgraded as a national key relic under special preservation in 2006.

First built in the Liao Dynasty, the Central, South and North seas used to be one water area. During the Jin Dynasty, the Palace of Great Peace was built here. In the Yuan Dynasty, the lakes were enclosed as part of the Imperial Palace. When Emperor Yongle rebuilt the Imperial Palace, he extended the walls to enclose both the former Yuan palace and gardens to the west. In the Ming and Qing, the area became known as the Western Gardens.

The three lakes were divided by bridges. The Centipede Bridge divides the South Sea from the Central Sea and the Golden Tortoise Jade Rainbow Bridge divides the Central Sea from the North Sea.

The Central and South Seas cover an area of 100 hectares, of which the Seas occupies 47 hectares. Most of the existing buildings date back to the Qing Dynasty. On the South Sea are the Precious Moon Tower and the Sea Terrace Island. On the Central Sea are the Tower of Vermilion Light, the Plantain Garden and the Pavilion of Clouds on the Water.

翔鸾阁
The Xiangluan Pavilion

涵元门
The Hanyuan Gate

蓬莱阁
The Penglai Pavilion

蓬莱阁，又名香宸殿，位于涵元
殿南面，重楼重檐，阁前立一"木
化石"，高约2.6米，两旁各有小
亭一座。

菊香书屋
The Chamber for Reading the Classics

颐年堂
The Yinian Hall

丰泽园
The Fengze Garden

# 景山
# Jingshan Hill

景山位于景山前街北侧，北京的市中心、城市南北中轴线的中心点，也是封建时代北京城的最高点。2001年被公布为全国重点文物保护单位。

元代，景山是皇帝的御苑。明成祖朱棣迁都北京，按照传统风水说法，紫禁城之北当有山，故将挖掘紫禁城筒子河和太液池南海所取出的泥土在此堆积成山，于四周种花植树，并在山之东北隅修了以寿皇殿为主的殿亭楼馆。其山取名为"万岁山"，据说山下曾堆放过煤，故又称"煤山"，同时，山下豢养成群的鹤、鹿，以寓长寿。明代在山上曾建有6个亭子，清初毁去。清顺治十二年(1655年)将万岁山改名景山。乾隆十五年(1750年)在山之五峰上各建一亭。次年，以供奉皇帝祖先神像的寿皇殿不应偏于一隅，而把寿皇殿移建到山后的中轴线上。

景山坐北朝南，南北约220米、东西约400米、占地面积约230050平方米，有东、南、西三座园门。景山主要建筑分布于山前、山后和山上三个部分。

山前部分，主体建筑由绮望楼及其附属建筑组成。绮望楼坐北朝南，在景山门北侧，位于中轴线上，依山而筑，此处原是清代乾隆年间办官学时，供奉孔子牌位的地方。

山后部分，主体建筑由寿皇殿、观德殿、永思殿以及牌楼、碑亭、神厨、神库等组成。寿皇殿始建于明代，清乾隆十四年(1749年)又进行重修，仿太庙建造，不仅规模宏伟，辉煌肃穆，而且自成一个完整的建筑格局。寿皇殿过去是供奉清朝历代皇帝神像的处所，现为北京市少年宫。永思殿和观德殿位于寿皇殿东，则是清代帝后停灵之所，现为少年宫图书馆。

山上部分，景山由5座山峰组成，主峰43米，是中轴线的中心点，也是全城最高点。山上的5座亭子布局对称，分别建于各个山峰的山脊线上，自东向西为观妙亭、周赏亭、万春亭、富览亭、辑芳亭。5座亭中原各有一尊铜佛像，分别代表酸、苦、甘、辛、咸五味神灵，1900年均被八国联军掠走。

景山东麓，原有一棵老槐树，是明崇祯皇帝的自缢处。

Jingshan Hill the highest point during feudal times is located in the center of Beijing on the north to south axis. It was listed as a national key relic under special preservation in 2001.

Jingshan used to be a private garden of the imperial families in the Yuan Dynasty. After Emperor Yongle transferred the capital to Beijing, there should be a mountain at the north of the Forbidden City according to traditional geomantic omen. Therefore, the earth excavated to make the moat of the Forbidden City was piled up into a hill called Longevity Hill, and flowers and trees grew in the grounds. The Hall of Imperial Longevity was built on the northeast of the hill. Coal was once heaped around the foot of the hill, it was therefore also known as Coal Hill. At the same time, there were a lot of cranes and deer within this compound. It was renamed Jingshan in the 12th year of the reign of Emperor Shunzhi of the Qing Dynasty (1655 A.D.) and five pavilions were built on the five ridges of Jingshan in the 15th year of the reign of Emperor Qianlong (1750 A.D.).

Jingshan is about 220 meters long from the north to the south, and about 400 meters wide from the west to the east, covering an area of 230,050 square meters. The principal structures in the park are the Qiwang Tower, five pavilions, the Hall of Imperial Longevity, the Hall of Everlasting Memory, the Hall of Morals Observation etc.

景山南门
The Southern Gate of Jingshan Hill

绮望楼
The Qiwang Tower

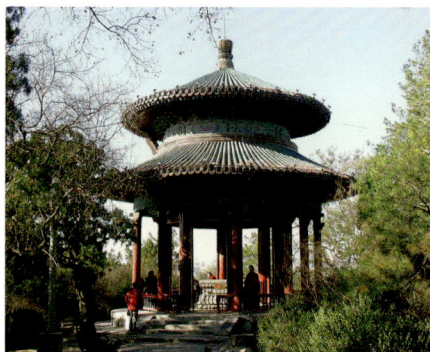

辑芳亭
The Harmonious Fragrance Pavilion

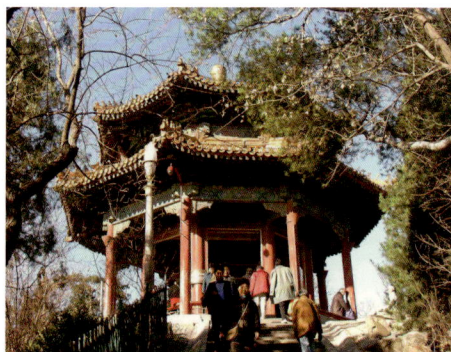

周赏亭
The Surrounding View Pavilion

万春亭
The Everlasting Spring Pavilion

景山全貌
The Jingshan Hill

富览亭
The Panoramic View Pavilion

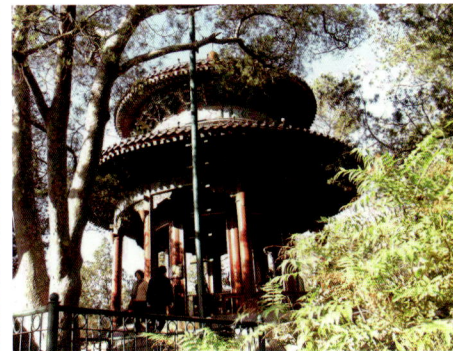

观妙亭
The Wonderful View Pavilion

寿皇殿
The Hall of Imperial Longevity

寿皇殿建筑群
The Hall of Imperial Longevity

吉祥阁
The Auspicious Pavilion

八角亭
The Octagonal Pavilion

# 颐和园
# The Summer Palace

  颐和园位于北京的西北郊，为"三山五园"之一，是我国现存最完整，规模最大的皇家园林，在世界古典园林中享有很高的声誉。1961年被公布为全国重点文物保护单位。

  颐和园原名清漪园。金代曾在此建立"西山八院"之一的"金山行宫"。明代在此修建了圆静寺和好山园。清乾隆十五年(1750年)，在圆静寺旧址建成大报恩延寿寺，次年，将瓮山改名万寿山，并将疏浚后的西湖(即金代瓮山泊)改名为昆明湖。咸丰十年(1860年)，英法联军抢劫了清漪园。光绪十二年(1886年)，慈禧挪用海军经费和其他银两，在清漪园的废墟上重新建园，光绪十四年(1888年)，改园名为"颐和园"。光绪二十六年(1900年)，颐和园再遭八国联军洗劫，光绪二十八年(1902年)，慈禧又动用巨款对颐和园前山再次进行修复。

  颐和园占地约290公顷，其中水面约占3/4，主要由万寿山和昆明湖等组成，北侧依山，南面临水。园内宫殿、寺庙、园林建筑布局，可概括为三大区域：政治活动区、居住区和游览区。

  政治活动区以东宫门内的仁寿殿为中心，是清末慈禧太后和光绪帝从事内政、外交政治活动的主要场所。这里虽是威严的政治区，但宫殿采用了灰瓦卷棚顶，院中又点缀以山石松柏，并建有花台，使之具有浓厚的园林气息。

  居住区以玉澜堂、乐寿堂、宜芸馆为主体，并以乐寿堂为中心，北依万寿山，南临昆明湖，3组院落由50余间曲折的游廊相连接，给人以舒适、和谐、宽敞的感觉，这里是慈禧太后、光绪帝及后妃居住的地方。

  游览区是全园的精华，融山水、建筑、花木为一体，是统治者的游憩之处，分前山、后山和昆明湖三个部分。

  颐和园继承了我国历代皇家园林的传统，又大量汲取了江南私家园林的造园艺术精华，兼有北方山川宏阔的气势和江南水乡婉约清丽的风韵，并蓄帝王宫室的富丽堂皇、民居的精巧别致和宗教庙宇的庄严肃穆，典型地反映了中国皇家园林特有的精神追求，代表了中国皇家园林修造的最高水平，有"皇家园林博物馆"之称，在中外园林史上有极高的地位。

颐和园晨曦
The Summer Palace in the Morning

Situated in the northwestern suburbs of Beijing, the Summer Palace is China's largest and best preserved imperial garden, known as one of the *Three Hills and Five Gardens*. It was listed by the State Council as a national key relic under special preservation in 1961.

The Summer Palace was originally called the Garden of Clear Ripples. When the Jin Dynasty made Beijing its capital, it built the Golden Hill Palace, one of eight gardens in the Western Hills on the present site of the Summer Palace. In 1750, Emperor Qianlong, to celebrate his mother's birthday, ordered the Garden of Clear Ripples to be built. In 1751, he changed the name of the hill to Longevity Hill. The whole project took 15 years and was completed in the 29th year of the reign of Emperor Qianlong (1764 A.D.). In the 10th year of the reign of Emperor Xianfeng (1860 A.D.), the Anglo-French Allied Forces robbed of the Garden of Clear Ripples and burned down most of the buildings. In the 12th year of the reign of Emperor Guangxu (1886 A.D.), Empress Dowager Cixi had it restored with the funds intended for the development of the navy and renamed it the Summer Palace. In 1900, it was again plundered by the Eight-Power Allied Forces. In 1902, Cixi spent a fabulous sum of money to have the palace reconstructed a second time.

万寿山全貌
The Longevity Hill

Composed mainly of the Longevity Hill and Kunming Lake, the Summer Palace covers an area of about 290 hectares, three quarters of which is water. The garden has three main areas: the political area, the living area and the sightseeing area.

The main hall in the political area is the Hall of Benevolence and Longevity. In the last years of the Qing Dynasty, Empress Dowager Cixi and Emperor Guangxu took charge of state affairs and received officials and foreign envoys here. However, the roofs of the buildings were covered with plain bluish grey tiles and the courtyards were embellished with evergreens, rocks and flower beds.

The living area is composed of three large courtyards, focusing respectively on the Hall of Jade Ripples, the Hall of Happiness and Longevity and the Yiyun Hall.

The sightseeing area, the major part of the Summer Palace, in harmony with landscapes, constructions and flowers and trees, can be divided into the Front Hill, the Back Hill and the Kunming Lake.

东宫门铜狮
The Bronze Lion in Front of the Eastern Gate

仁寿殿
The Hall of Benevolence and Longevity

仁寿殿内景
Interior of the Hall of Benevolence and Longevity

长廊内部梁架
Interior Framework of the Long Corridor

长廊
Long Corridor

长廊彩画·包公执法
Coloured Painting of the Long Corridor

长廊彩画·倒拔垂杨柳
Coloured Painting of the Long Corridor

长廊彩画·智套宗保
Coloured Painting of the Long Corridor

长廊彩画·岳母刺字
Coloured Painting of the Long Corridor

长廊彩画·三难新郎
Coloured Painting of the Long
Corridor

长廊彩画·牛郎织女
Coloured Painting of the Long
Corridor

长廊彩画·风雪山神庙
Coloured Painting of the Long
Corridor

长廊彩画·归田乐
Coloured Painting of the Long
Corridor

石舫
The Marble Boat

长廊是中国园林中最长的游廊
The Longest Corridor in Chinese
Gardens

乐寿堂外景
Exterior of the Hall of
Happiness and Longevity

乐寿堂内景
Interior of the Hall of Happiness
and Longevity

德和园大戏楼
The Opera Theater in the Garden
of Virtuous Harmony

慈禧看戏的宝座
The Throne for Empress Dowager Cixi to Watch Beijing Operas

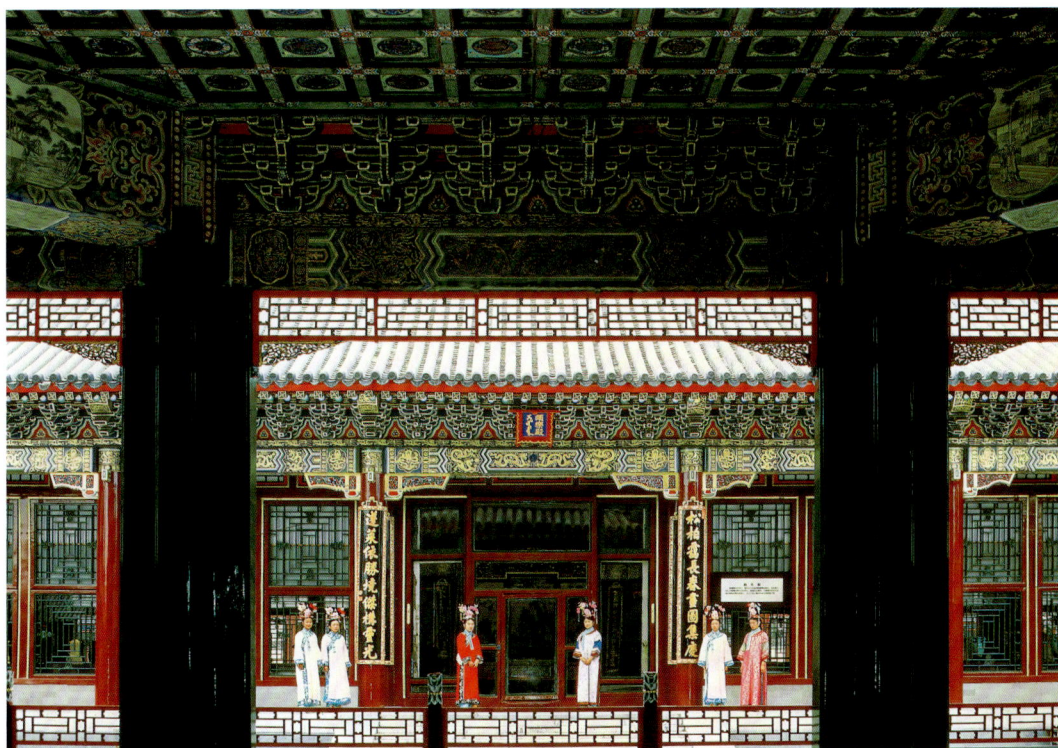

颐乐殿
The Hall of Natured Joy

排云殿
The Hall of Dispelling Clouds

万寿山前山
The Front Hill of the
Longevity Hill

佛香阁内景
Interior of the Tower of
Buddhist Incense

佛香阁
The Tower of Buddhist
Incense

后山中部宏伟壮丽的藏传佛教建筑
Magnificent Tibetan Buddhist Buildings in the Central
Section of the Back Hill

众香界和智慧海
Zhongxiangjie and the Hall of the
Sea of Wisdom

四大部洲一隅
A Corner of the Four Continents

多宝琉璃塔
The Glazed Pagoda

谐趣园雪景
The Perspective of Snow in the
Garden of Harmonious Interest

知鱼桥
The Zhiyu Bridge

谐趣园鸟瞰
The Bird View of the Garden of
Harmonious Interest

谐趣园一角
A Corner of the Garden of Harmonious Interest

苏州街冬景
The Perspective of Snow in
Suzhou Street

苏州街
Suzhou Street

十七孔桥
The Seventeen-Arch Bridge

铜牛
The Bronze Ox

十七孔桥
The Seventeen-Arch Bridge

西堤
The Western Bank

远眺玉泉山
The Distant View of the Jade
Spring Hill

学府衙署

# ACADEMYS AND GOVERNMENT OFFICES

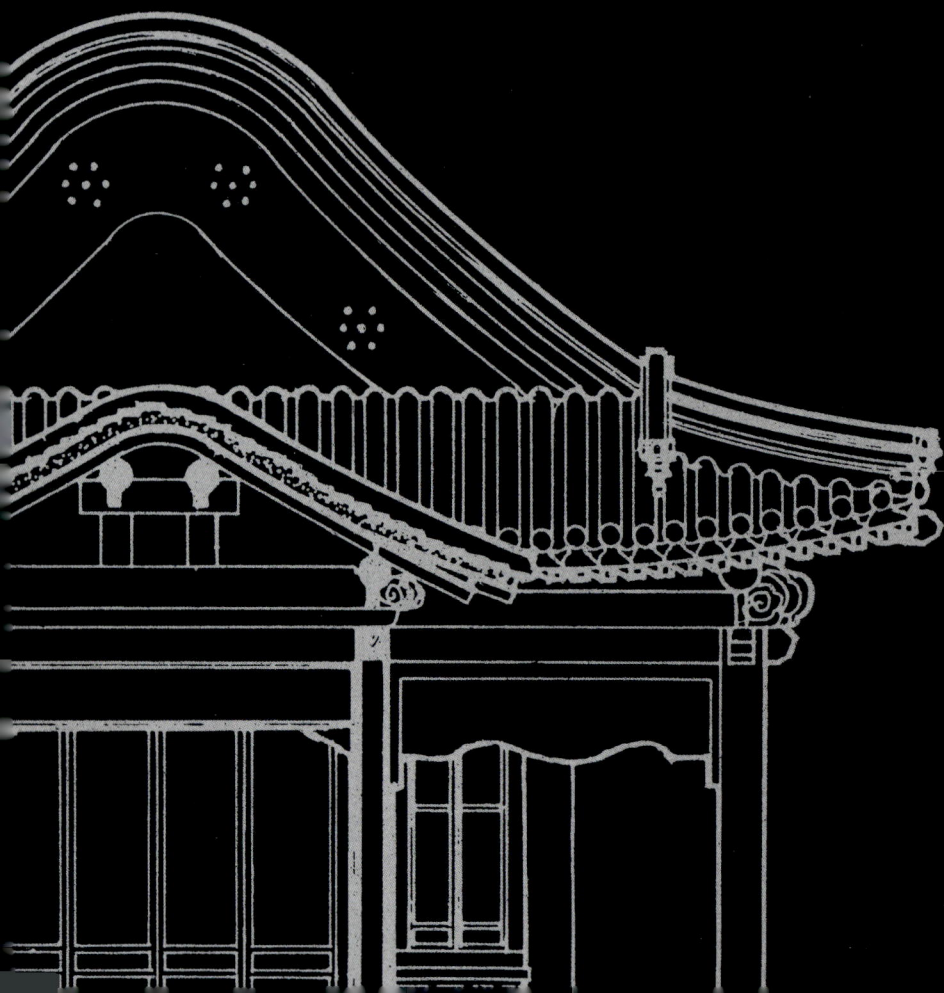

# 北京古代的学府衙署建筑

作为中华民族先祖的诞生地和华夏文明的发祥地之一，北京在漫漫的历史长河中，其所处的地理位置、历史环境，始终处于多种文化交汇的前沿，尤其是元明清以来，北京成为政治中心，科教文化更是出现了前所未有的兴旺，学府衙署建筑也随之大量兴建。

## 一、北京古代的学府建筑

### 1.元代以前

根据史籍记载，北京至少在西周时期的燕国已经开办了学校，"学在官府"，学校都是官办，私人办学尚未出现。当时的官学分为设在都城内的"国学"和按行政区划设立的地方官学——"乡学"。春秋战国时期，在思想、文化方面出现百家争鸣，在其影响下，官学日渐衰败，私学开始出现。关于这一时期学校建筑的情况，文献记述甚少。

自秦、两汉，历经魏晋南北朝、隋唐到五代，今北京地区地处边陲，其教育状况随着朝代更迭、国家政策方针的改变发生着变化，时而兴盛，时而衰败。但总体来说，无论官学，还是私学都还是不断发展的。这些时期，北京地区学校的建筑情况记述较少。

辽金时期，随着北方游牧民族的崛起，北京的政治、经济地位发生了巨大的变化，逐渐成为全国的政治经济中心，与此同时，北京也变为全国的教育中心。辽代，在南京(今北京)设立国子学，北京地区所属各州县均设地方官学，州称州学，县称县学。金代，北京升为国都——金中都，是金王朝的政治、文化中心，共有三所中央官学：国子学、太学、女真国子学。此外还设立一些特殊的专门学校，如学习天文历法的司天台天文学等。北京地区的地方官学有大兴府学、女真府学、诸州学、大兴府医学等。辽金时期，北京地区的私学主要是教授儿童识字的乡塾。有关这一时期学校的建筑情况亦记述甚少。

### 2.元明清时期

元朝立国建都，北京成为名副其实的全国教育中心。

到明清时期，自明成祖迁都北京后，北京既是全国的政治中心，也是文化中心，北京古代学校教育迎来极盛时期，学校建筑也得到极大发展。

1233年，元代在燕京设立学校，这就是最早的国子学，国子学是元代全国最高学府。至元六年(1269年)，元世祖诏令正式设学，学址先在南城，至元二十四年(1287年)迁到现在国子监的位置。同时设立管理机构国子监，专门负责教育工作。国子学校舍分为上、中、下三等，每等各两斋，东西相向。

明朝时，国子监亦是全国的最高学府，入监学生称为监生，其来源有三：一是皇帝亲自指派的勋戚、官僚子弟，称"官生"；二是由各府、州、县地方学校选拔的高材生，称"民生"；三是外国留学生。

清承明制，国子监仍是全国最高教育行政机关和最高学府，入监的学生有各省荐举的贡生、输银纳捐的监生以及不经考选的荫生。

北京国子监坐北朝南，按"左庙右学"之制，由三进院落组成，主要建筑按中轴线对称布局。国子监自建成之日起，屡经修葺、扩建，功能不断健全完善，特别是乾隆年间的扩建，最终形成今天看到的规模。

在国子监建成近500年后，于乾隆四十九年(1784年)建成辟雍殿。按周代的礼制，国学设在天子的国都中，称为辟雍，一般认为是天子承师问道，行礼乐，宣教化的地方。"辟雍"二字含有扬善戒恶、明和天下的意思。在清乾隆之前，辟雍只存在于传说中，是一种有水有殿，又有寓明和鉴戒之意的独特建筑。乾隆皇帝及辟雍的设计者，根据前人的解释，加上自己的见解，创造性地建成了世界上唯一的一座辟雍圜水工程，使千百年来流传的"辟雍"建筑展示在世人面前。

辟雍四面开门，方形大殿建在圆形的池水中央，四面有石桥通达，外圆内方的布局是有讲究的：天圆地方，池圆象征德圆，殿方象征行方，是体天体之撰，立规矩之极也；四周环以水、达以桥，是附会"水圆如璧"的说法，同时以水为界限，用于节制观者。从建筑上说，辟雍是装饰性和实用性的完美结合：屋角向上，屋腰下沉，在下大雨的时候，能使屋面雨水流冲较远，不致溅入走廊；同时，体现了曲线美；独具匠心的铜制镏金宝顶，起到结构与装饰的双重作用；建筑整体色彩富丽，气势雄伟的抹角梁架使建筑结构十分合理，内部空间宽敞，符合教学功能的需要。辟雍建成200多年来，抗拒了多次地震的摇撼，它的质量经受住了严峻的考验。

元代的中央官学除国子学外，另设有回回国子学和女真国子学。明清两朝，中央官学除国子监外，明朝亦设有专业学校、太医院、钦天监、四译馆等在负责各自日常工作的同时，兼有培养医学、天文历法、翻译人才的责任；同时还设立武学，培养能征善战的军事人才。清朝，八旗分别在本旗地界内设官学一所，属国子监管辖；并设景山官学和咸安宫学，学习满汉书，属内务府管理。有关这些学校的建筑情况，文献资料记载甚少。

元代在今北京地区，设路、州、县、社各级官学。明代北京地区的地方官学十分发达，顺天府所属各州均设学，县以下乡里又有社学。清代北京的地方官学大体沿袭明制，设有顺天府学。

顺天府学是明清两朝北京地区最高的地方官学，现有建筑分为东西两路，是右庙左学之制。在西路第一进院中有一椭圆形泮池，泮池是文庙的特有建筑。按周代礼制，国学设于天子和诸侯的国都中，《礼记·王制》："太学在郊，天子曰辟雍，诸侯曰泮宫"。诸侯所立国学，等级低于天子的辟雍，因此水只环半圈，成半璧状，称为"泮池"。自秦代废除诸侯分封之制，后世遂在州县官学的文庙中比拟诸侯泮宫兴建泮池。

### 3.北京的书院

"书院"在中国历史上存在了约1200年，它产生于唐代，兴于宋代，废止于清代，在中国古代教育史上占有重要地位。北京地区在唐五代时期是北方军事重镇，胡汉杂处，文化相对落后；五代以后，包括北京在内的燕云十六州被割让给辽国；宋与辽、宋与金对峙时期，文化交流又

受到阻碍。因此，北京地区错过了唐、宋书院的兴起、兴盛的潮流，元以前的北京地区不见书院的记载。

元代是北京设立书院的开始，北京地区共建立三所书院：大都城区的太极书院、昌平的谏议书院、房山的文靖书院，其建筑今已无存。

明代各种私学较发达，除义学、各类塾学、冬学、乡学等初级学校外，还有书院。明代书院的发展经历了曲折的过程，经历了多次毁书院的举动，明代北京地区书院见于记载的共有6所：京师首善书院、通州通惠及双鹤书院、密云白檀书院，还有承袭元代的昌平谏议书院、房山文靖书院，其中较著名的是首善书院。据记载，首善书院有讲堂、后堂各三间，供有孔子牌位，藏有经世典籍。

清代，北京地区的私学较明代更发达，除有义学和塾学外，主要还有书院。清代允许在官府的严密控制下建立书院，清朝在北京地区设立的书院数量，为历代之最，遍及京城和所有州县，较著名的是京师金台书院。金台书院主体建筑系三进四合院，房屋开间较大。据记载，共有房64间。

## 二、北京地区学府建筑的特点

首先，学校建筑中最主要的是有大讲堂、斋舍，建筑以讲堂为中心。如国子监的辟雍，就位于国子监建筑群的中心位置；国子监中东西六堂，作为教室，每堂共为十一间。再如金台书院，其东西文场均面阔十间。

其次，庙学结合，庙是学的信仰中心，学是庙的存在依据。所有的学宫均设有文庙，两者并列，布局方式多为文庙在左，学宫在右，即所谓"左庙右学"，这种建筑形式是按照礼制规定设立的。如国子监、孔庙，就是遵循"左庙右学"的礼制，孔庙即文庙在东侧(左)，国子监是太学，在西侧(右)。文庙是祭祀孔子，建有大成门、大成殿，同时配祀先贤、哲人，包括崇圣祠、启圣祠、名宦祠、乡贤祠、文昌祠等。这些建筑并非所有文庙均有，而是根据其规模大小而有所取舍，府学、州学的孔庙规模较大，县学、书院的文庙规模较小。如顺天府学，它是地方官学其

中的乡贤祠、名宦祠、大成殿，其建筑规模、等级要比附属于国子监的北京孔庙中的小得多。

## 三、北京古代的衙署建筑

北京作为五朝古都，全国的政治中心，曾建有许多为封建王朝服务的衙署建筑，如掌管观察天象，推算节气历法的钦天监、举行科举考试的贡院、中央八部的衙署等。只可惜随着封建王朝的覆亡，许多衙署建筑也随之湮灭无存。北京现存的衙署建筑中比较著名的是观象台和升平署。

### 1.观象台

古观象台是中国著名的古代天文台，始设于元代，原名"司天台"。明初攻克北京时毁于战火，残存的天文仪器被运往南京保存。明正统七年(1442年)时重建此台，名"观

星台"，同时复制了一套仪器，并修复了台下的紫微殿、漏房等建筑。清代时又增添了一些仪器，并更名为"观象台"。古观象台是由高台砖砌建，是中国古代天文学光辉成就的历史见证，它以观测历史久远、仪器设备保存完好而享誉世界。

### 2.升平署

升平署是清代掌管宫廷戏曲演出活动的机构，始于康熙年间，隶属内务府管辖。初是令太监在此学戏，后又兼管召选宫外艺人进宫当差演戏或充做教习的事务。直到宣统三年(1911年)，升平署前后历时共有一个半世纪之久，为研究我国的戏剧史保存、提供了大量珍贵的实物资料。升平署戏楼院是保存较好的一组建筑物。

# Ancient Academys and Government Offices Architecture in Beijing

As Beijing became a political center, the science, education and culture showed unprecedented flourish, and a large number of college buildings were built up.

## 1. Ancient Academy Architecture in Beijing

### (1) Before the Yuan Dynasty

According to the historical records, schools in Beijing were open as early as the State of Yan in the Western Zhou Dynasty, and all those schools were officially run, and private schools had not emerged yet. The official schools of the time were divided into two parts: national schools established within the city, and village schools, local official ones established in administrative regions. In the Spring and Autumn and Warring States Periods, many schools of thoughts and culture contended. Under such circumstance, the official schools eclipsed day by day, and the private schools began to emerge. School architecture in this period is seldom mentioned in historical literature.

During the period from the Qin, Eastern Han and Western Han Dynasties, the Wei, Jin, and Northern and Southern Dynasties, to the Sui and Tang Dynasties, the present-day Beijing belonged to the border area, and the education status in this area witnessed rise and fall, and varied with the changes of the dynasties and the alteration of the national policies. On the whole, either official schools or private schools developed constantly. So, few of them were recorded in historical literature.

In the Liao and Jin Dynasties, with the rise of northern nomadic people, Beijing's political and economic status witnessed great changes, and gradually became a national political and economic center, and also the national education center. In the Liao Dynasty, the Imperial College was established in Nanjing (present-day Beijing), and local official schools were all established in Beijing's states or counties, those schools established in states were called state colleges, and those set up in counties called county

colleges. In the Jin Dynasty, Beijing was upgraded as a national capital named Jin Zhongdu, a political and cultural center in the Jin Dynasty. There were altogether 3 central official schools, namely, the Imperial College, the National College, and the Jurchen people's Imperial College. In addition, some specialized schools were established, such as the school of astronomy which taught astronomical and calendric knowledge. The local official schools in Beijing included the Daxingfu College, the Jurchenfu College, the Zhuzhou College, and the Daxing Medical School. In the Liao and Jin Dynasties, the private schools in Beijing were mainly countryside schools. There are few records on the school architectures in this period.

### (2) The Yuan, Ming, and Qing Dynasties

In the Yuan Dynasty, Beijing became the real national education center. In the Ming and Qing Dynasties, since the Emperor Mingchengzu moved the capital to Beijing, the city became the political and cultural center in the country. The ancient school education in Beijing reached its peak, and the school architecture was developed greatly.

In 1233, college was established in Yanjing in the Yuan Dynasty, which was the earliest and highest level of school in the Yuan Dynasty. In the 6th year of the Zhiyuan reign (1269 A.D.), the Emperor Yuan Shizu gave imperial edict to establish the school first at the southern city. The school was relocated to the place where the present-day Imperial College is located in the 24th year of the Zhiyuan reign. The Imperial College was also established as administrative organization, and specialized in charging of education. The houses of the Imperiad College are divided into three levels, each level having two rooms, and the eastern rooms and western rooms facing together.

In the Ming Dynasty, the Imperial College was the highest level of school. The education mechanism formed in the Ming Dynasty was reserved in the Qing Dynasty. The Imperial College remained to be the administrative organ and highest level of school in the whole country. The students of the school included Gongseng (Second-

Degree Scholars) recommended by each province, Jiansheng who were enrolled through contributing silver and donation, and MengSheng who were enrolled without need to take entrance examination.

The Imperial College, facing south, consists of three courtyards one behind another in accordance with the system of "Temple Left and College Right". The main buildings in the College were arranged symmetrically alongside the central axis.

Shuntianfu College was the highest level of local official college in the Ming and Qing Dynasties. The existing architecture is divided into the eastern and the western axes, following the system of "Temple Left and College Right". In the first courtyard on the western axis is an elliptic pool named Panchi which is the characteristic construction of the temples. According to the ritual system in the Zhou Dynasty, the national colleges were established in the capital city where the emperor and feudal princes resided, as the *Liji Wangzhi*, (Book of Rites, Royal Regulations) says, "for the imperial college located in the outskirts, the emperors called it Piyong, and the feudal princes called it Pangong". The national colleges established by feudal princes have lower grade than the Piyong Hall for emperors, therefore, the pool before the former colleges take on semi-circular shape, and was called Panchi. Since the Qin Dynasty abolished the subinfeudation system, the offspring thus copied the model of Pangong to construct Panchi in the Temple of Confucius in the state or county schools.

### (3) Academies in Beijing

The Yuan Dynasty marked the start of establishing academies in Beijing. There were altogether three academies in Beijing, namely, the Taiji Academy in Dadu, the Jianyi Academy in Changping area, and the Wenjing Academy in Fangshan area, none of the three left at present.

There are a total of 6 academies in Beijing on record, i.e. the Shoushan Academy in the capital, the Tonghui

Academy and the Shuanghe Academy in Tongzhou area, the Baitan Academy in Miyun, and the Jianyi Academy in Changping area which followed the style of the Yuan Dynasty, and the Wenjing Academy in Fangshan, of which the Shoushan Academy was more famous. According to records, the Shoushan Academy has the lecture hall and back hall, with a tablet of Confucius, and ancient books and records.

The Qing Dynasty witnessed the largest number of academies established in Beijing, and these schools spread all over the capital city and all the states and counties, of which the Jintai Academy was more famous.

## 2. Characteristics of Academy Architecture in Beijing

Firstly, the school architecture consisted mainly of big lecture halls and houses, with lecture hall as the center of the buildings. For example, the Piyong Hall of the Imperial College was located in the central position in the building group.

Secondly, temples served as the belief center of colleges, while colleges were the base for existence of temples. In all the colleges erected temples, and the two types of buildings stood side by side. The pattern was such that the temples stood in the left side and the colleges in the right, i.e. the so-called "Temple Left College Right". This architectural style was formed according to the ritual system. For example, the Imperial College and the Temple of Confucius followed the ritual system of "Temple Left College Right". The Temple of Confucius, i.e. Temple of Literature, was located in the eastern side, while the Imperial College which was the national university, was located in the western side. The Temple of Literature was used for worshipping Confucius. There were the Great Accomplishment Gate and the Great Accomplishment Hall and also ancestral temples for worshipping sages and philosophers, including Chongsheng Ancestral Temple, Qisheng Ancestral Temple, Minghuan Ancestral Temple,

Xiangxian Ancestral Temple, Wenchang Ancestral Temple and so on. Ancestral temples were not built in all temples of literature, and it depended on the scale of buildings. The Temples of Confucius in the Fuxue and Zhouxue were relatively large, while those temples were smaller in Xianxue and academies.

# 3. Ancient Architecture of Government Offices in Beijing

Beijing which is the ancient capital of five dynasties and national political center, had many ancient architecture of government offices serving feudal dynasties, such as Directorate of Astronomy which observed astronomical phenomena, and calculated solar terms and calendar, Examination Office which was in charge of imperial examinations, and six central government departments. Unfortunately, with the fall of feudal dynasties, many government office architecture have disappeared without trace. Among the famous government office architecture existing in Beijing are the Ancient Beijing Observatory and the Shengpingshu.

## (1) The Ancient Beijing Observatory

The Ancient Beijing Observatory, originally called Sitiantai, is a famous astronomical observatory established in the Yuan Dynasty. It was destroyed during the war of attacking Beijing in the early years of the Ming Dynasty. The remaining astronomical instruments were carried to Nanjing for preservation. A new observatory was built in the 7th year of the reign of Emperor Zhengtong of the Ming Dynasty (1442 A.D.), called Constellation Observatory. Meanwhile, a set of instruments were also reproduced, and the Ziwei Hall, Hourglass House and other constructions under the observatory were renovated. During the Qing Dynasty, the observatory was equipped with some more instruments, and was renamed as Ancient Observatory. The observatory, built of bricks, is an evidence of the splendid achievement of Chinese ancient astronomy. It is famous in the world for its long history and well-preserved instruments.

## (2) The Shengpingshu

The Shengpingshu served as an organization that was in charge of opera performing for the royal court in the Qing Dynasty. It was first built during the reign of Emperor Kangxi, and was affiliated to the Imperial Household Department. At the beginning, eunuchs were ordered to study drama performance here, and later, it summoned actors outside the court to make performances or serve as instructors. The Shengpingshu remained in existence for one and a half century until the 3rd year of the reign of Emperor Xuantong (1911 A.D.). It preserves and provides a wealth of precious physical materials for the research of Chinese drama history. The Opera Theater in Shengpingshu is among the well-preserved architecture.

# 国子监
## Guozijian (the Imperial College)

国子监位于北京市东城区国子监街15号，是元、明、清三代国家设立的最高学府。1961年被公布为全国重点文物保护单位。

北京国子监始建于元大德十年（1306年），明代初期，定都南京，一度将北京国子监改称为北京府学，明成祖迁都北京后，永乐二年（1404年）又改为国子监。民国时期和中华人民共和国成立以后，国子监均进行过不同程度的修缮。1956年辟为首都图书馆，2004年图书馆搬出，国子监成为博物馆。

国子监坐北朝南，按"左庙右学"之制，东邻孔庙，由三进院落组成，占地两万多平方米。院内柏树参天，肃穆静谧，主要建筑全部集中在一条中轴线上，自南而北依次为集贤门、太学门、琉璃牌坊、辟雍、彝伦堂和敬一亭，附属建筑围绕各自的主体建筑分布，这些主次建筑共同构成国子监的主体。其中辟雍所在的第二进院落是国子监最大的院落，亦是全监主要建筑的集中地，院内分别建有辟雍、东西六堂、博士厅、绳愆厅、点簿厅以及牌楼和钟鼓亭等建筑，左右对称，排列有序，布局合理，环境优雅。

国子监的每组院落均有围墙环绕，这种做法不仅满足了功能使用上的需要，而且也使其区域划分更为合理，等级的区别和互不干扰成为国子监建筑的特点之一。在国子监的外围建有较高的围墙，这不仅符合中国传统建筑的规范和制度，同时也使得国子监因与外界的隔绝而备显庄重和神圣。

The Imperial College at No.15 Guozijian Street in Dongcheng District was the highest institution of learning in the Yuan, Ming and Qing Dynasties. It was listed as a national key relic under special preservation in 1961.

It was first built in the 10th year of the Dade reign of the Yuan Dynasty (1306 A.D.) and renamed Fuxue in Bejing (the highest educational organ). After Emperor Yongle transferred the capital to Beijing, it was renamed Guozijian again.

Located to the west of the Confucian Temple, Guozijian, which faces south, occupies about 20,000 square meters of land and comprises three courtyards one behind another. Dotted with ancient cypresses, the principal structures are Jixianmen (the front gate), Taixuemen (the second gate), the Glazed Archway, the Biyong Hall, the Yilun Hall (the Hall of Sacrifice of Ethics) and the Jingyi Pavilion all standing in the central axis. The second courtyard where Biyong stands is the largest and all the main buildings are laid out there. In the courtyard are Biyong, Six Halls to the east and west, the Boshi Hall, the Shengyan Hall, the Dianbo Hall, the archway and bell and drum towers.

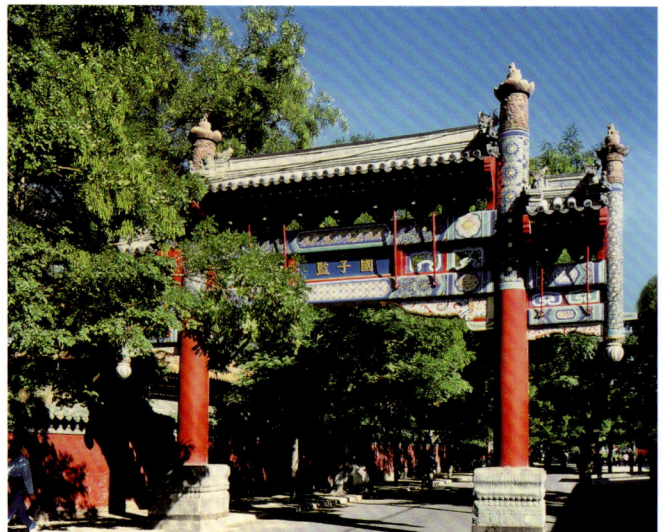

国子监街牌楼
The Archway in Guozijian Street

琉璃牌坊
The Glazed Archway

太学匾额
The Board Inscribed with
"Taixue"

从琉璃牌坊看辟雍
Looking at the Biyong Hall from
the Glazed Archway

辟雍泮水
Pond of the Biyong Hall

辟雍是国子监建筑群的核心建筑，清乾隆四十八年（1783年）开始筹建，清乾隆四十九年（1784年）落成，是清代帝王讲学的场所。自清康熙以后，每位皇帝即位，照例要到国子监讲学一次，称作临雍。清代国子监的辟雍，大殿建在高大的方形石基之上，殿为重檐四角攒尖顶，面阔与进深均为三间，四周建有围廊（副阶周匝），四面无墙，均装槅扇，以便讲学时敞开。

辟雍建在一座圆形水池的中央，池水上有4座石桥，与辟雍的东、西、南、北四门相通，连接内外。池岸和石桥上建有汉白玉石护栏，护岸壁上有4个龙头。圜水围绕辟雍，这种建筑形式称为"辟雍泮水"，是参考了古代关于辟雍的记载而建。整个建筑布局外圆内方，

寓意天圆地方、传流教化。

在辟雍的东、西两侧各有房33间，即东西六堂。东侧从南到北有崇志堂、诚心堂和率性堂；西侧自南而北为广业堂、正义堂、修道堂，每堂共为11间，是专供监生学习的场所，相当于现代的教室。六堂建筑外廊较大，可供监生在廊下活动。每座堂的正中檐下部位都悬挂有各堂的名称，建筑规范整齐，成为国子监中院建筑群的主要组成部分。

辟雍
The Biyong Hall

辟雍隔扇门
Partition Doors of Rooms of
the Biyong Hall

辟雍门饰
The Decoration on Doors of the
Biyong Hall

鼓亭
The Drum Pavilion

彝伦堂
The Yilun Hall

在辟雍之北，明代永乐年间在元代崇文阁旧址上重建时改名为彝伦堂，为国子监藏书的地方。辟雍未建成之前，皇帝在此举行临雍典礼，后为传经授业的主要场所。建筑坐北朝南，面阔七间，进深九檩，后出抱厦，单檐悬山顶。彝伦堂前建有宽大的月台，又称灵台或者露台、平台等，它是国子监召集监生列班点名之处。在月台的东南角设有日晷一部，是古代依据日形测定时辰的仪器，又称日表。西南的汉白玉石须弥座上置有赵孟頫所书的《乐毅论》石刻。

鸟瞰国子监
The Bird View of the Imperial College

# 顺天府学
# Shuntianfu College

顺天府学位于北京市东城区府学胡同65号，是明、清两代顺天府的官学。1984年被公布为北京市文物保护单位。

顺天府学原为元代太和观(即报恩寺)旧址。明洪武初年，此地为大兴县学。明永乐元年(1403年)，升北平为顺天府，在京师设国子监，此处不得再设县学，故改成顺天府学，大兴、宛平县学附于府学。永乐九年(1411年)建明伦堂东西斋舍，永乐十二年(1414年)建大成殿，又建学舍于明伦堂后。宣德年间重建大成殿，并配以东西两庑及戟门。万历年间，督学商为正将文丞相祠从学舍西侧迁至东侧，后来督学杨四知在文丞相祠的东北侧建尊经阁，在东南侧建文昌祠。

顺天府学坐北朝南，现有建筑分为东西两路。西路建筑有两进院落，主要为祠堂建筑，正门为棂星门，第一进院棂星门北面为椭圆形泮池，池上架三座石桥。院子西侧为乡贤祠三间；东侧为名宦祠三间。泮池往北为大成门，面阔三间。穿过大成门，进入第二进院落，大成门北面为大成殿，是西路建筑的主体建筑，面阔五间，庑殿顶。东路建筑即为顺天府学堂部分，大门三间。大门北面为二门三间。左右官厅、祠殿各三间，做教室用。仪门在二门北面，门内有明伦堂五间，两侧为斋舍。明伦堂之东为奎星阁，为六角二层阁楼。

顺天府学的所有建筑中，大成殿是清初遗构，乡贤祠、名宦祠是原状修复，府学二门尚是原物，其余均为2000年原址复建。

大门
The Front Gate

Located at No.65 Fuxue Hutong in Dongcheng District, the Shuntianfu College was an official institution of learning in the Ming and Qing Dynasties. It was listed as a Beijing's relic under preservation in 1984.

Originally the Taihe Taoist Temple was built in this place in the Yuan Dynasty. In the early years of the reign of Emperor Hongwu of the Ming Dynasty, an institution of learning in Daxing County was built here. Because Beiping was upgraded as Shuntianfu in the 1st year of the reign of Emperor Yongle (1403 A.D.), this place was changed to Shuntianfu College, which was renovated many times during the Ming Dynasty.

The Shuntianfu College, which faces south, is divided into the eastern and the western axes. The buildings on the western axis comprise two courtyards one behind another. In the first courtyard lies the Lingxing Gate, to the north of which is a pond shaped like an ellipse. Three stone bridges span over the water. On the western side is the Xiangxian Ancestral Temple with 3 bays wide and on the eastern side is the Minghuan Ancestral Temple with 3 bays wide. To the north of the pond stands the Great Accomplishment Gate, which is 3 bays wide. Through the gate, you will enter into the second courtyard. The main hall, the Great Accomplishment Hall is 5 bays wide with a hip roof.

大成门
The Great Accomplishment Gate

# 金台书院
# Jintai Academy

　　金台书院位于北京市崇文区崇文门外东晓市街203号，是清代康熙年间建立的一所义学。1984年被公布为北京市文物保护单位。

　　此地原是降清明将洪承畴的赐园——洪庄。康熙三十九年(1700年)，京兆尹钱晋锡在宛平、大兴分设义学，收孤寒生童就学，其中大兴义学赁屋于洪庄。后来，宛平的义学合并到洪庄来，改称首善义学，但还只是租赁洪庄里一些房屋。后来，洪氏之孙奕沔献地办学。首善义学在原有基础上增建校舍，扩大规模。乾隆十五年(1750年)正式改名为金台书院，隶属顺天府，学员主要是京师和各省准备参加会试、殿试的举人和贡生，但顺天府的童生亦可就读。

　　道光二十二年(1842年)和光绪五年(1879年)对其进行过二次规模较大的重修。当时书院主体是三进四合院式院落，布局井然有序。光绪年间重修后的建筑计有大门一座，垂花门一座，东门一座，大堂三间，官厅六间，朱子堂七间，南罩房五间，东西文场各十间且带回廊，东西厢房十间，加上厨房、马栅等共计六十四间。

　　光绪三十一年(1905年)废除了延续千余年的科举制度，推行学校教育，金台书院停办，其旧址改为顺直学堂，民国期间改为公立第十六小学，后几易其名，校舍却无变动，1973年改为崇文区东晓市小学，现名崇文区金台小学。

　　金台书院的主要建筑现在均保留下来，近年又经过修缮，面貌焕然一新，外显文雅、庄重。现存文物有乾隆四十九年(1784年)《金台书院记》石刻一方，嵌于门洞东壁。金台书院自开创义学至今，已300年历史，并且始终为学校使用。

Located at No.203 Dongxiaoshi Street outside Chongwenmen, the Jintai Academy used to be an volunteer school during the reign of Emperor Kangxi of the Qing Dynasty. Now, it is a Beijing's relic under preservation in 1984.

The place used to be Hong Chengchou's Residence, called Hongzhuang. Hong Chengchou was at first the Ming general. Then, he surrendered to the Qing Dynasty. During the reign of Emperor Kangxi, Qian Jinxi ran a volunteer school here. The school was formally renamed Jintai Academy in the 15th year of the reign of Emperor Qianlong (1750 A.D.), being subject to Shuntianfu. Then, it was renovated twice. Because the imperial examinations were abolished in the 31st year of the reign of Emperor Guangxu (1905 A.D.), the Jintai Academy closed down. It was changed into the public No. 16 Primary School in the Republic of China and was renamed Dongxiaoshi Primary School in Chongwen District in 1973.

二门
The Second Gate

正殿
The Main Hall

庑房
The Hip Room

# 皇史宬
# Huangshicheng (the Imperial Archives)

皇史宬位于北京市东城区南池子大街136号，原皇城内东南角，靠近明代的南内。1982年被公布为全国重点文物保护单位。

皇史宬始建于明嘉靖十三年(1534年)，原用来贮藏明代历朝皇帝的《宝训》、《实录》的正本。以后《永乐大典》的副本也保存在这里。清代移走明代的《实录》，用来储存清代的《实录》、《圣训》、《玉谍》等。清嘉庆十二年(1807年)曾重修，但其规模及主体建筑基本上未变。

皇史宬一组建筑有南北中轴线，共二进院落，前院是狭长的通道，外门在东西两侧。院内中央北面为正门，门内为一长方形庭院，北面正中一座面阔九间、单檐庑殿顶的无梁殿就是贮存档案的皇史宬。

The Imperial Archives is situated at No.136 Nanchizi Street in Dongcheng District, the southeastern corner of the Imperial City. It was listed as a national key relic under special preservation in 1982.

First built in the 13th year of the reign of Emperor Jiajing of the Ming Dynasty (1534 A.D.), it was rebuilt in the 12th year of the reign of Emperor Jiaqing of the Qing Dynasty (1807 A.D.), maintaining the original scale and design.

Used to store important historical documents, the Imperial Archives comprise two courtyards and the main buildings are situated on a north-south axis.

皇史宬匾额
The Board Inscribed with
"the Imperial Archives"

皇史宬正殿
The Main Hall of the Imperial Archives

皇史宬全部用砖石建成，主要原因显然是为了防火，但同时也为了附会古代国家藏书处为"金匮石室"的记载。殿身建在两米多高的石台基上，正面开5个券门作入口，各门均为两层，外层石门，内层木门；山面各开一个方窗。殿前有月台，正中台阶间有云龙御路。殿内据明末记载有20个石台，上贮金匮，到清乾隆时已改为两个大台子，是清末改建的。

皇史宬外观为仿木构建筑。墙身由灰色磨砖砌成，檐下的柱头、额枋、斗拱、椽子都是砖石所做，但目视几乎与木构无殊。为了适应砖石材的特点，斗拱出挑较短，用条石上刻万拱架在拱头上，出檐也较短，外观比一般木构建筑显得厚重。

皇史宬外观体型宏大，色调雅致，在灰色砖墙身和黄琉璃瓦顶之间有青绿点金的斗拱额枋彩画，衬着下面的汉白玉石栏杆、台基，给人以更为明朗、庄重的感觉。

金匮
Jinkui

# 升平署戏楼
# Opera Theater in Shengpingshu

升平署戏楼位于北京市西城区西长安街1号。1984年被公布为北京市文物保护单位。

升平署是清代掌管宫廷戏曲演出活动的机构，称南府，始于清康熙年间。清乾隆五年（1740年）设南府于南花园（在今南长街南口），令太监在此排戏，隶属内务府管辖。为区别西华门内之内务府，故称位于南花园的分府为南府。乾隆十六年（1751年），下谕选苏州艺人进宫当差，命名为外学，令住景山，仍属南府管辖。原习艺太监命名为内学。道光七年（1827年）将外学撤销，艺人回原籍。又将十番学并入中和乐内，增设档案房，改南府为升平署，仍主持宫内演出事务。嗣后又兼管召选宫外艺人进宫当差演戏或充做教习的事务。直到宣统三年（1911年），升平署

前后历时共有一个半世纪之久。1912年，中南海改为总统府，升平署物品移到景山。升平署珍藏的剧本、档案、戏衣、道具、剧照等，至今保存在故宫博物院内，成为我国戏剧史上珍贵的实物资料。

升平署戏楼是保存较好的一组建筑物，建筑面积约200平方米，这座戏楼在一组四合院中，坐南朝北，戏楼北面的北房前出轩，适合帝后观赏演出。戏楼台基高0.8米,宽12米，进深11米。台口4柱，四角各有角柱3根，场门原为城门样式，寓"出将"之意，下场门原为宫门形式，寓"入相"之意。后墙上下之间的正中位置，原有大宫门，可排仪仗。戏楼为两层，下层台顶中央有活动天花板，上下有木楼梯，戏楼南侧有3间扮戏房。

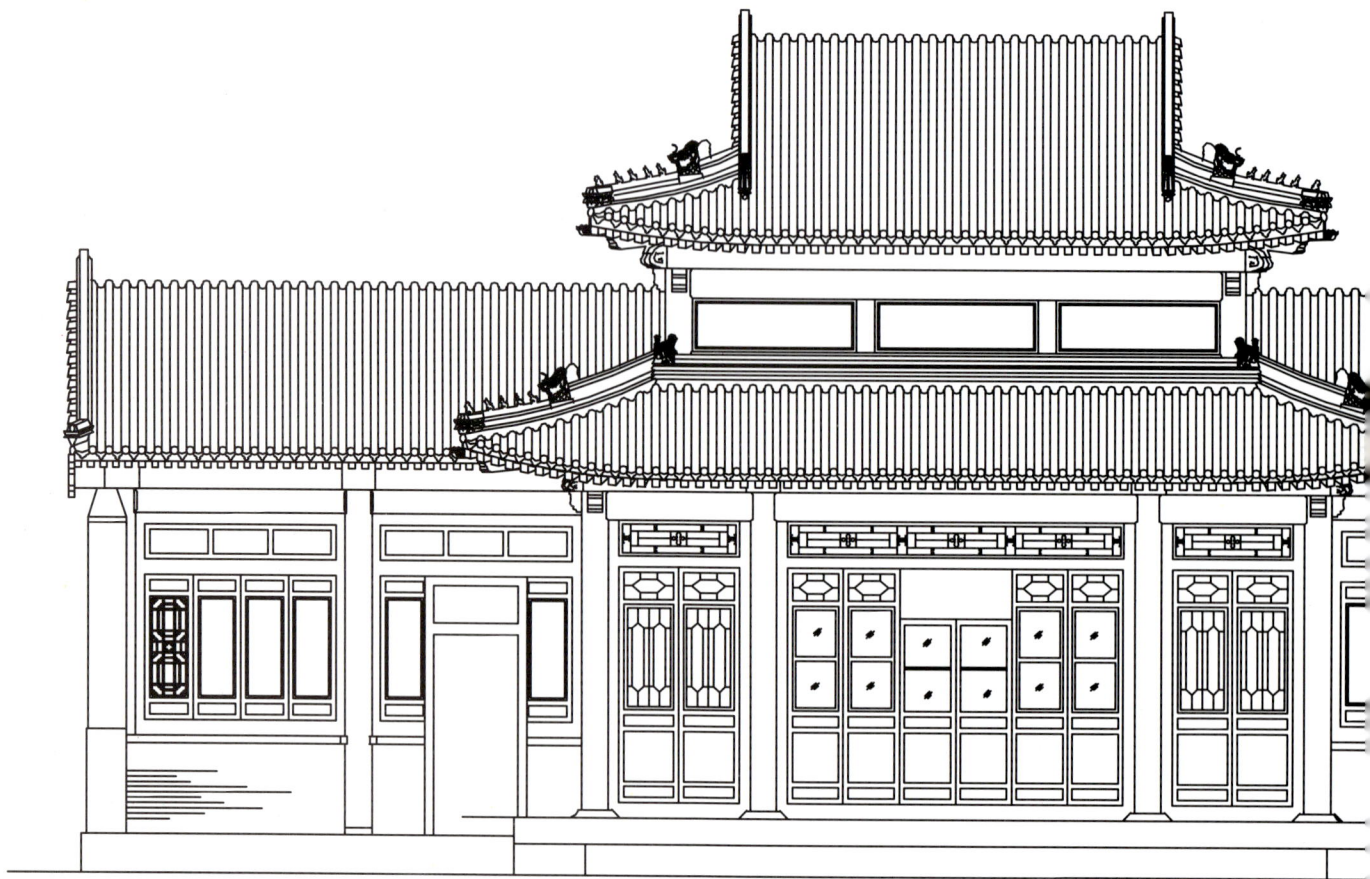

升平署戏楼北立面图
The Elevation of the Opera Theater in Shengpingshu

Located at No. 1 Xichang'an Street in Xicheng District, it was listed as a Beijing's relic under preservation in 1984.

The Shengpingshu served as an organization that was in charge of opera performing in the Qing Dynasty, also known as Nanfu.

The Opera Theater in Shengpingshu was first built during the reign of Emperor Kangxi. Covering an area of about 200 square meters, it, which faces north, comprises one quadrangle. The base of the opera theater is 12 meters wide, 11 meters deep and 0.8 meters high. There are three pillars on each corner of the opera theater. It has two stories.

戏楼北立面
The Elevation of
the Opera Theater

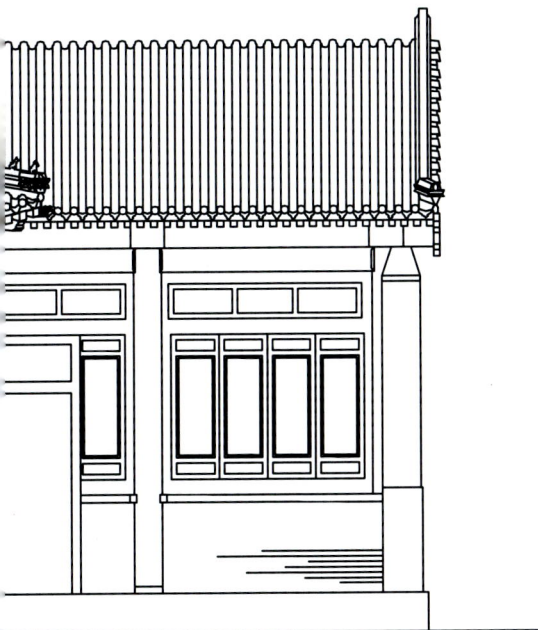

观戏房
The Chamber for Watching Operas

# 古观象台
# The Ancient Beijing Observatory

古观象台位于北京市东城区建国门立交桥西南侧，是世界上现存最古老的天文台之一，同时也是明清两代天文观测的中心，它以建筑完整、仪器配套齐全、历史悠久而闻名于世。1982年被公布为全国重点文物保护单位。

北京地区的天文台始于金代的候台，元代在今建国门观象台北侧建司天台。明正统四年至七年（1439－1442年），在元大都城墙东南角楼旧址改筑为台体，建观星台，并在城墙下建紫微殿等房屋，正统十一年（1446年）又增建了晷影堂，此时观星台和其附属建筑群已颇具规模，基本形成了今天的布局。清代因观天仪器增加，观象台先后共向东拓宽了8米。

古观象台为砖砌高台建筑，上窄下宽，平面呈凸形，台高为14.25米，1980年修缮时将城台内掏空，辟为二层展厅，8件大型铜制仪器陈列在台上南、西、北三面。台体西侧是以紫微殿、东西厢房和晷影堂为主的附属建筑群，建于1442－1446年，清乾隆九年（1744年）重修。紫微殿面阔五间，东西附耳房各三间，殿前有东西厢房各五间。庭院南侧为大门三间，两侧为耳房三间，又接顺山房各三间。从平面上看，该庭院由三条轴线组成，中路（大门、耳房、紫微殿）为礼仪部分，西路（西侧顺山房、西厢房、西耳房）为管理用房，东路（东侧顺山房、东厢房、东耳房）为测量用房。庭院东南角另有晷影堂三间，原有铜圭铜表，是测量夏至、冬至日射角的场所。

古观象台是我国也是世界上使用年代最久，古代天文仪器数量最多而又保存最完整的历史文物。从明正统年间到1929年连续观测近500年，在世界上现存的观象台中，保持着同一地点上连续观测最久的历史记录。古观象台不仅进行天文观测，也进行气象观测，它保存了自清雍正二年（1724年）至光绪二十八年（1902年）共180年中每天的气象资料，是世界上现存最早的气象观测记录。

Located at the southwestern side of Jianguomen Crossroad in Dongcheng District, the Ancient Beijing Observatory is one of the oldest observatories in the world. At the same time, this place served as an astronomy observation center in the Ming and Qing Dynasties. It was listed as a national key relic under special preservation in 1982.

In the Jin Dynasty, the rulers ordered to build the Chief Astronomer's Observatory in the Beijing area. Then, the Yuan Dynasty built an observatory named Sitiantai just north of the site of the present-day the Ancient Beijing Observatory. The Constellation Observatory was constructed on the site of the Southeastern Corner Tower of Yuan Dadu from the 4th year to the 7th year of the reign of Emperor Zhengtong of the Ming Dynasty (1439-1442 A.D.). The Ziwei Hall and other buildings were built under the tower. The Sundial Shadow Hall was added in the 11th year of the reign of Emperor Zhengtong (1446 A.D.). It was during that period that the Constellation Observatory and its affiliated buildings took on present scale and layout.

The Observatory is an elevated brick-built platform structure with broad base and narrow top. The height from the bottom to the top is 14.25 meters. Eight bronze astronomical instruments stand on the south, west and north of this platform, to the west of which are affiliated buildings, such as the Ziwei Hall and the Sundial Shadow Hall, built from 1442 to 1446. They were renovated in the 9th year of the reign of Emperor Qianlong of the Qing Dynasty. The Ziwei Hall is 5 bays wide with side rooms with 3 bays attached to either side. On the eastern and western sides, there are wing rooms with 5 bays wide respectively. To the south of the courtyard stands the Front Gate with 3 bays with side rooms with 3 bays attached to either side. The courtyard is divided into the middle, the eastern, and the western axes.

天体仪
The Celestial Globe

地平经仪
The Azimuth Theodolite

古观象台全貌
The Ancient Observatory

府邸宅院

# MANSIONS
# AND RESIDENCES

# 北京的住宅建筑

北京作为五朝古都，长期居住着皇室贵族及士大夫，这种阶层对居住环境的高要求，从各个方面促进了北京住宅建筑的发展与完善。再加上北京地区的地理位置、气候条件、风俗民情等因素，共同促成了北京独具特色的建筑文化。

## 一、北京的住宅简述

北京的传统住宅俗称"四合院"，是北方庭院式住宅的代表。四合式住宅的优点，是有一个隔离于外界的内部小环境。在形式上，它体现了住宅所要求的私密性、安全性和独立性。在功能上，它满足了封建社会父权统治、男尊女卑、主仆有别的家庭伦理秩序的要求。因此，四合院建筑也成为全国许多地区共同采用的住宅形式。北京现在的四合院形式最直接肇始于元代。元大都城的规划产生了胡同与胡同之间的四合院住宅，经过明清两朝，这种住宅形式进一步得到发展，于是"北京四合院"成了北京住宅的代名词。

明清北京四合院与元代时的四合院有较明显的变异，院落的"工"字形平面布局消失，占地面积也明显减少。从元代四合院遗址来看，其前院更宽敞，而明清时的四合院则是前院面积较小而后院面积较大，分布更合理。此外，清代时出现了一个特殊的现象，就是满汉分居，内城只准满人居住，回、汉等其他民族只能居住在外城。而且随着城市的发展，逐渐形成了城东部主要以商贾居住为特色，西部主要以王府、公卿宅邸为主，外城居住平民为特点的居住格局。北京的住宅一般分为大型的王府住宅和中型的公卿豪富宅邸以及小型的民居。小型四合院一般都由大门、影壁、正房、厢房、耳房、走廊组成。一般分为前后两院。两院之间由二门(中门)连通，前院用作门房、客厅等，后院为主人内宅则非请勿入。其中位于宅院中轴线上的堂屋，规模最大，建筑质量最高，为全宅的核心部分。大中型四合院如王府、公卿府邸、达官富商宅院等则通常由多进和多路四合院组成，它的规制、格局承袭了古代宫室建筑的特点，尤其是王府，均为前堂后寝的多进复合式

宅院，多设有客厅、饭厅、寝室、花园、后罩房、佣人房、车马房等，院落层叠，气派非凡。

## 二、北京的清代王府建筑

### 1.北京清代王府制度

北京最大型的住宅要算王府了。由于明代采取分封制，所以明代的王府多分布于全国各地，就算是史书上记载的王府井一带的十王邸也早已不存在了，所以北京现存的王府均为清代所建。在清代，王府是封建社会等级最高的贵族府邸。王府属于皇家财产，由内务府统一管理，一旦撤爵，就会被收归内务府。

清代诸王有世袭罔替和世袭递降两种，世袭罔替者俗称"铁帽子王"，指其爵位可以世代沿袭，该爵位始终不变，袭爵者如因罪夺爵，可选同宗承继。世袭递降者是指后世子孙的爵位要比其父辈低一等。例如父亲是亲王，儿子就袭封郡王，孙子则袭封贝勒。如果最初被封为亲王的，其后代降到镇国公就不再降了；若最初被封为郡王，则其后代降到辅国公就不再降了。当爵位与其府邸不符时，该王府就会被收回，需另择他处居住。因此，这种王府便存在一府多主的变化。如果一座王府中出了皇帝，该王府就成了"潜龙邸"，要改建成宫殿，不能再居住，原府主人则由内务府另赐新府。王府根据其爵位的高低有明确的规定。《清会典》中规定："凡府第各颁其制。亲王府制，正门五间，启门三间，缭以崇垣，基高三尺。正殿七间，基高四尺五寸。翼楼各九间，前墀环护石阑，台基七尺二寸。后殿五间，基高二尺。后殿七间，基高二尺五寸。后楼七间，基高尺有八寸。共屋五重。正殿设座，基高一尺五寸，广十一尺，后列屏三，高八尺，绘金云龙。凡正门殿寝，均覆绿琉璃瓦，脊安吻兽，门柱饰以五彩金云龙纹，禁雕刻龙首。压脊七种，门钉纵九横七。楼房旁庑，均用筒瓦。其府库、仓廪、厨厩及典司执事之屋，分列左右，皆板瓦，黑油门柱。亲王世子府制，正门五间，启门三，缭以重垣，基高二尺五寸。正殿五间，基高三尺五寸。翼楼各五间，前墀护以石阑，台基高四尺五寸。后楼

五间，基高一尺四寸。共屋五重。殿不设屏座。梁栋绘金彩花卉、四爪云蟒。金钉、压脊各减亲王七分之二。余与亲王同。郡王府制亦如之。贝勒府制，基高二尺，正门一重，启门一。堂屋五重，各广五间。筒瓦压脊，门柱红青油漆、梁栋贴金、彩画花草。余与郡王府同。贝子府，基高二尺，正门一重，堂屋四重，各广五间，脊用望兽。余与贝勒府同"[《清会典》(光绪朝)卷五十八、工部]。由上面记载可知王府较一般民居建筑等级高、规模大，一般由中、东、西三路组成。中轴线上有府门、正殿、寝殿、后楼等组成。

## 2.北京清代王府的建筑特点

王府大门一般不直接临街而建，门外围有一座宽大的庭院，类似皇宫的宫前广场，庭院北面是王府正门，门前两侧置石狮一对。南面正中为倒座房，面阔三至五间，是王府中的管事处、回事房。进出府邸都从东西两侧的阿斯门(满语"翅膀"之意，俗称雁翅门)。因阿斯门临街，门外还要设下马桩及行马，所以王府前庭常常将府前街拦腰截断，一般官吏和市民至王府前均须绕行。如和亲王府地处东城区张自忠路东口，西阿斯门正对大街，就使该路至此需绕过王府前庭才能通至东四北大街。也有一些王府因地段关系或其他原因不设前庭院，大门直接沿街布置，但在大门前的街道对面建有巨大影壁，如克勤郡王府大门对面就建有长达22米的起脊式四岔软心影壁。王府大门多为五间形式，多为单檐歇山顶，檐下施斗拱，绘龙锦彩画，亲王府大门可覆绿琉璃瓦。大门两侧多建有带抱厦的旁门，为平时的出入口，前庭院倒座房及左右阿斯门间用低矮的辅助用房相连接。

进入大门是外朝的主体的庭院，是整座王府建筑的最大的院落，中有高起的甬路和台基，正面为王府的正殿(俗称银安殿)，亲王府七间，郡王府五间。正殿是整座王府中规模最大、等级最高的建筑，多为歇山顶，檐下施斗拱和龙锦彩画。正殿可设座，座后置金云龙彩屏。是亲王或郡王召见属员和举行各种大典的场所。庭院用楼庑环绕，完

全采用对称的布局，翼楼间数也有明确规定，亲王府九间，郡王府五间。所有建筑都施不同等级彩画。主体庭院面积宏阔，建筑高大，给人以恢宏富丽之感。大殿后为寝门，寝门是前朝后寝的分界线，相当于故宫乾清门的功能，多数为三间，门与两庑相连，围成封闭的内寝院。进入寝门，正面为寝殿，亲王府七间，郡王府五间。个别的府中寝殿前加建三至五间抱厦，两侧建有配殿、顺山房。这组建筑是亲王或郡王及其福晋起居之处。寝殿后面为王府的最后一重院落，正中为两层的后罩楼，亲王府九间，郡王府七间。自前庭院至后罩楼共五重院落，构成了王府建筑的主轴线。由此我们可以看出王府建筑就是一个皇宫的缩影，从其轴线布置主要建筑，府前广场，前朝后寝的格局等等都与皇宫的布局相似，只是规模上小很多。再考虑众多王爷的皇室身世，可以说王府是皇家文化的重要组成部分。

但是从现存的王府分析，王府主轴线上的门、殿、楼、寝都要严格按规制建造，不得逾制。因为逾制后轻则罚俸，重则夺爵治罪，如清初开国功臣八大"铁帽子王"之一的郑亲王就因擅自使用御路石而被夺爵，因此绝大多数的王府在规划建造时，往往在一处或数处低于朝廷的规定。如礼亲王，也是清初八大"铁帽子王"之一，府内的翼楼仅有7间。清同治、光绪年间的恭亲王奕䜣，因在辛酉政变中支持慈禧太后夺权有功，授议政王，并命王爵世袭，军政大权集于一身，但其府邸的大门仅有三间，正殿也仅五间。两侧无翼楼，只有五间配殿，比朝廷规制低得多。再如清代末期赫赫有名的"一府二帝"的醇亲王，老醇亲王奕譞和其子载沣的儿子相继承继皇位，是为光绪帝和宣统帝，后载沣曾任摄政王，其新府是在成亲王府的基址上花了16万两银子进行的大规模翻修。但府中翼楼也只有五间。其他一些王府低于规制的地方就更多了。

在北京的王府中还有少数是蒙古王府。一般来说，蒙古王爷自领蒙古地区的本旗属地，京城之内不赐府邸，只有极少数例外，如超勇亲王策凌府、僧格林沁的僧忠亲王府等，但这些王府不是按照清代宗室王府规制修建的，只

不过是由几个规模较大，质量较高的四合院组成而已。

### 3.北京清代王府的保存现状

据统计，目前北京尚存有王府19座，其中列为全国重点文物保护单位的有6座，为恭亲王府、醇亲王府、孚王府、雍亲王府(雍和宫)、淳亲王府(属"东交民巷使馆建筑群"之一)、和亲王府(清陆军部和海军部旧址)。北京市文物保护单位中有王府8座，为礼亲王府、庆王府、郑王府、恒亲王府、老睿亲王府(普度寺)、克勤郡王府、宁郡王府、循郡王府。区级文物保护单位中有王府4座，为醇亲王府南府、敬谨亲王府(清学部旧址)、惠亲王府、仪亲王府。

清代覆亡后北京的王府开始被破坏，至解放初期尚有六七十处保存较完整的。然而多年以后，北京城内的王府大部分已不复存在了。幸存的少数王府中有的也只保留下部分殿堂或院落，有的不少作为民房或改作他用。尽管一些王府已经被列为"文物保护单位"，但破坏并没有停止。只有恭亲王府、醇亲王府、孚王府等少数几个保存尚好，具备开放游览的条件。

## 三、北京中小型的住宅建筑

北京中小型四合院多是坐北朝南，由正房、倒座房(南房)和东西两侧的厢房围合而成的一个封闭性庭院，宅门一般开在院落的东南角。这是根据阳宅风水术所谓"坎宅巽门"的布局而设的。正房坐北，大门开在东南方;正好落在八卦的坎位与巽位上，寓意财源广进、家宅平安。有的房宅在街道南侧，大门朝北，仍要保持"坎宅巽门"的布局，就在庭院外侧加一条走廊，使走廊南端通入院内。在南北向街道两侧的住宅多把宅门开在临街一侧的偏南处。

小型四合院一般为一到两进，正房和厢房多为面阔三间，正房东西两侧或厢房南端一般可置耳房一至二间，在北京传统住宅建筑中相当普遍。这种小型四合院修建比较简单，宅门多为随墙门、如意门等，只占一间或半间。如意门往往在门楣及山墙墀头上，雕有精美的砖雕花饰，这也是旧京民宅中使用最为普遍、最具特色的大门形式。

中型院落多是官僚贵族、富商巨贾的住宅。它们大多是由多个院落纵向排列在一条中轴线上，南北贯通，可达四五进之多。一些较大的宅院在东西两侧还带有跨院，作为祠堂、花园、马号等附属用房。中型院落的宅门一般为广亮大门、金柱大门或蛮子门，这些大门前檐多饰有雀替、彩画，大门内外多做独立式一字影壁，有的还在大门山墙左右斜向外伸出两堵墙，称为八字影壁，扩大了门前空间，使门楼显得高大气派。大门内第一进院北面正中一般建有装饰华丽的垂花门，门内第二进院子北面正中为过厅，两侧为东西厢房。垂花门、过厅与厢房间用游廊相连接。游廊多为绿色梅花方柱，柱间做出精美的倒挂楣子、坐凳楣子以及各式各样的什锦窗，精致考究。以后各院形式基本与第二进相同。这类院落的室内外装修多数都很华丽，很多使用名贵的木料，做成隔扇门窗、碧纱橱、几腿罩、圆光罩等室内外装修。在庭院绿化上也很有特色，一般门口都种槐树，院内则种植些小型乔木、灌木之类，一则不会遮挡阳光，二则春华秋实，还能寓意吉祥富贵，如种植石榴(寓意多子多孙)、牡丹、海棠(寓意玉堂富贵)。庭院中还有一个特色摆设就是大鱼缸，北方干燥，四合院又都是木建筑，这样既可以调节小气候，又可以在必要之时用于救火，还可以观赏，可谓一举三得。四合院中还有众多类似的既有美好寓意又实用的陈设、装饰，已经形成了独具特色的四合院文化，很是耐人寻味。

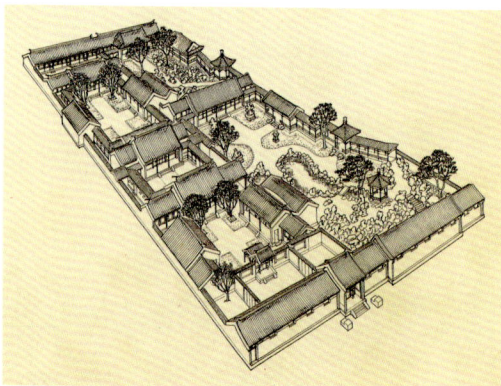

带花园的四合院布局示意图

# Beijing's Residential Architecture

Beijing, the ancient capital of five dynasties, had long been inhabited by royal aristocrats and scholar-bureaucrats who had high requirements on the settlement environment, thus, promoting the evolvement and improvement of the residential architecture in Beijing. This, combined with the factors such as geographical location, climate conditions, and customs, contributed to the unique architectural culture in Beijing.

## 1. Brief Introduction on Residences in Beijing

The traditional residence in Beijing, commonly called quadrangle, is a representative of the courtyard-style residences in northern China. The advantage of the quadrangles is that there is an inner small surrounding isolated with the outside world. In terms of form, the quadrangles reflect the requirements on residence, namely, privacy, security, and independence. In terms of function, they met the requirements of the family ethic orders such as paternity sovereignty, male superiority, hierachical difference between the master and servant etc. The quadrangles existing in Beijing started to be built in the Yuan Dynasty. The layout of the Yuan Dadu gave rise to Hutongs (small lanes) as well as the quadrangle residences between two Hutongs. This kind of residence was developed further in the Ming and Qing Dynasties.

Compared with those in the Yuan Dynasty, Beijing's quadrangles in the Ming and Qing Dynasties changed a lot, such as the disappearance of "工" shaped layout of the courtyard, and sharp decrease in the occupied area. The site of the quadrangle of the Yuan Dynasty showed a more spacious front courtyard, while that in the Ming and Qing Dynasties had a smaller area in the front courtyard but a larger area of the rear courtyard, with more reasonable arrangement. In addition, a particular phenomenon occurred in the Qing Dynasty was separate residence of the Manchu people and the Han people, i.e. the inner city was only for residence for the Manchu people, while the Han people and other ethic people were only allowed to reside in the outer city. The development of the Beijing city led to the formation of the following residence pattern: the merchants settled in the east of the city, the royal families mainly resided in the west, and the commoners lived in the outer city. The Beijing's residential buildings were divided into the large-scale mansions of princes, and medium-sized residences of senior officials and wealthy men, as well as small-sized residences of commoners. Small-sized quadrangles generally consisted of front gates, screen walls, principal rooms, wing rooms, side rooms, and corridors, and comprised two courtyards one behind another. In the front courtyard stood gatehouses and guestrooms, while in the back courtyard were the owner's inner chambers which should not be admitted without invitation. The principal rooms standing on the middle axis were the core part of the whole architecture with the largest scale and highest quality. Middle- or large-sized quadrangles, such as princes' mansions, senior officials' dwellings, the residences of government officials and rich merchants, comprised many courtyards one behind another and were divided into many axes. Their hierachical design and layout were inherited from the ancient palaces. In particular, princes' mansions were all complex residences with the pattern of "halls in the front and resting rooms at the back". They usually consisted of guestrooms, dining rooms, bedchambers, gardens, shielding rooms, servant rooms, the houses for carriages, and stables.

## 2. The Princes' Mansion Architecture in Beijing in the Qing Dynasty

### (1) The system in princes' mansions in Beijing in the Qing Dynasty

The largest residence in Beijing should be the princes' mansions. All the princes' mansions remaining in existence in Beijing were built in the Qing Dynasty. The princes' mansions were the highest-grade aristocratic residence in the feudal society. They belonged to the royal property, and were under the unified management of the Inner Affairs Department. Once the princes were removed of the titles of nobility, their mansions would be confiscated by the Inner Affairs Department.

Prince's mansions are larger and higher-grade than residences of common people. A prince's residence can be divided into the middle, the eastern and western axes. On the middle axis are the gate, the main hall, the sleeping hall, the posterior shielding storeyed building and so on. The front gate of a prince's residence is 5 bays wide with 3 pairs of doors at the center. The main hall is 7 bays wide with the throne and the folding screen inside, flanked by wing buildings with 9 bays wide. The roofs of all the halls are covered with green glazed tiles, while the roofs of other rooms with grey pantiles. The gate of a commandary prince's residence is also 5 bays wide with 3 pairs of doors at the center.

## (2) The architectural characteristics of the princes' mansions in Beijing in the Qing Dynasty

The mansion gate was not built nearby the street, and outside the gate is a spacious courtyard, similar with the front plaza in the imperial palace, and north of the courtyard is the front gate of the princes' mansion flanked by a pair of stone lions. The gate is 5 bays wide with gable-and-hip roofs. The roof beam and the interior of the pavilion are painted with dragons. The roof of the gate is covered with green glazed tiles. Thus, it can be seen that princes' mansion construction is an epitome of imperial palace. Expect a small scale, the princes' mansions have similar layout with the imperial palace in terms of the arrangement of major constructions along the axis, plaza in front mansion, and the pattern. Considering the royal linkage of a large number of princes, the mansions constitute an important part of the royal culture.

Entering the front gate, you will find the main courtyard of the outer palace. The courtyard is the largest one in the whole prince's mansion, with a raised gravel path and podium in the middle. On the front side is the main hall (also called Yin'an Hall) of the mansion. There are usually seven rooms in main hall in princes' mansions and five rooms in main hall in prefecture rulers' residences. The main hall is the largest construction with highest rank among the whole mansion. It is normally built with gable and hip roof, and with wood brackets on the eaves and dragon coloured paintings under the eaves. A seat is placed in the main hall, and behind the seat is a golden dragon color screen. The hall was used by princes or prefecture rulers to summon their subordinates or to hold grand ceremonies. The courtyard, encircled by many rooms, is arranged symmetrically. There is also a definite limit on the number of wing rooms, that is, nine in princes' mansions and five in prefecture rulers' residences. All architectures were applied with coloured paintings with different grades.

Behind the main hall is the sleeping gate. The gate serves as a boundary between the halls in the front and the resting rooms at the back. It has similar function with the Gate of Heavenly Purity in the Palace Museum. Entering the sleeping gate, one will find a sleeping hall in the front side. There are seven rooms in sleeping hall in princes' mansions and five in prefecture rulers' residences. In some mansions, three to five annexes were built before the sleeping hall.

Behind the Sleeping hall is the last courtyard in the mansion. In the central front side is a two-storied shielding building. There are nine rooms in shielding building in princes' mansions, and five rooms in prefecture rulers' residences. There are altogether five courtyards from the front to the rear, constituting the main axis of the mansion architecture. We can see from the above description that the mansion architecture is an epitome of an imperial palace. Except smaller in size, the mansion's layout is similar with the imperial palace in the following aspects: main buildings arranged based on its axis, the square before the mansion, and the pattern of "hall in the front and resting rooms at the back". In consideration of the royal lineage of many princes, it can be said that the mansions are an important part of loyal culture.

## 3. Middle and Small Sized Residential Architecture in Beijing

The middle or small sized quadrangle, which faces south, is compound with the principal room, the reversibly-set room, the wing rooms on both sides so that a square inner courtyard is formed at the center. The main entrance is at the southeastern corner. The small sized quadrangle comprises one or two courtyards one behind another. The side rooms with 1 or 2 bays wide are attached to either side of the principal room with 3 bays wide. On the eastern and western sides are the wing rooms with 3 bays wide

respectively.

The greening of the courtyards is characteristic: the pagoda trees grow at the entrance, and small arbor or shrubberies grow in the courtyard, serving the following functions: do not sheltering from the sunlight, and blooming in the spring and bearing fruits in the autumn, having metaphorical meanings of auspiciousness, the richness and honor, for example, the common pomegranate indicates many sons and many grandchildren, and Chinese flowering crabapple represents richness and honor. Another characteristic of the courtyard is to place a big fish tank in it. Since the climate in the northern China is dry, and the quadrangles are wooden buildings, the fish tank served for three functions: view appreciation, adjustment of the microclimate, and also fire fighting. There were many similar furnishings and decorations which are practical and have beautiful meanings, bringing out the unique and thought-provoking the quadrangles culture.

# 恭亲王府
# Prince Gong's Residence

恭亲王府位于北京市西城区前海西街17号，是北京现存最典型、保存最完整的清代王府和花园。1982年被公布为全国重点文物保护单位。

这里原是乾隆朝时权臣和珅的宅邸。嘉庆四年（1799年）和珅获罪，府邸被转赐给庆亲王永璘，改为庆王府。咸丰二年（1852年），咸丰帝将此宅转赐给六弟恭亲王奕訢，因而改称恭亲王府。光绪二十四年（1898年），奕訢死后，王爵由其裔孙溥伟继承，直至1937年溥伟将王府及花园转卖给辅仁大学。

恭亲王府由府邸和花园两部分组成。府邸部分占地3.1公顷，分为东、中、西三路，均为多进四合院组成。府邸之北为占地2.6公顷的王府花园，名萃锦园。园内有布局精妙的曲廊、亭榭、山石、水池及花木等。

恭亲王府现存正门两座，南向门内原有银安殿及东西配殿已焚毁，只剩下基础部分。后殿面阔五间，为恭亲王府内举行萨满教祭祀的场所。中轴线上的建筑均为硬山调大脊，前后出廊，绿琉璃筒瓦屋面，梁枋绘旋子彩画，配殿则用灰筒瓦顶。

东路的前院已毁，中院正厅五间，名多福轩，为会客厅，后院正厅五间，名乐道堂，为奕訢的起居之所。西路为居住区，中院正厅五间，名葆光室。后院南有一垂花门，一殿一卷形式，内悬"天香庭院"匾额。北侧正厅为锡晋斋。在三路院落北端环抱着长160余米，面阔五十间的通脊二层后罩楼，东曰瞻霁楼，西曰宝约楼。

恭亲王府花园名萃锦园，位于后罩楼北侧，布局可大致分为中、东、西三路。中路正门为西式砖雕拱券门，内有一蝙蝠形的水池，名福河，池后为安善堂。两侧游廊分别连接东西各三间配房。其后又有一水池，池后由假山堆叠出石洞，洞内嵌康熙御笔的"福"字碑。假山之上为邀月台，建有三间敞厅，两侧有爬山廊连接东西配房。中路最后的建筑名为养云精舍。东路南端有一六角形流杯亭，名沁秋亭。过一垂花门北侧为大戏楼，这是北京王府中仅存的一座戏楼，价值很高。

西路外缘砌有一段带雉堞的城墙，名榆关。榆关内东有妙香亭和秋水山房。西侧为益智斋三间。北部有一大水池，池中水榭三间，名诗画舫。花园中的池水由后海引入。在清代除皇家苑囿以外，仅有三处准许引活水入宅，恭亲王府名列其中，可见当年奕訢的尊贵地位。

石狮
The Stone Lion

王府大门
The Front Gate

Located at No.17 Qianhai Xijie in Xicheng District, the Prince Gong's Residence is now the most typical and best-preserved of the princes' mansions and gardens of the Qing Dynasty in Beijing. It was listed as a national key relic under special preservation in 1982.

Formerly, it was the private residence of Heshen, a favorite minister of Qianlong reign. In 1799, Emperor Jiaqing had Heshen executed. He raised the status of Heshen's residence to that of a prince's, and bestowed it to Yonglin, Prince Qing and renamed Prince Qing's Residence. In 1852, the mansion was bestowed to Yixin, Prince Gong by Emperor Xianfeng, hence the name.

The Prince Gong's Residence consists of two parts: the living quarters and the garden. The living quarters, containing an area of 3.1 hectares, are divided into the middle, the eastern and western axes. On the middle axis the Yin'an Hall and the halls on each side have been destroyed by fire, only the basis of which remained. The rear Jiale Hall is a sacred place for Shamanism. On the eastern axis, the first courtyard has been destroyed. In the second courtyard, the principal room is 5 bays wide, called Duofuxuan. In the third courtyard, the principal room is 5 bays wide, flanked by wing rooms on both sides. On the western axis, the second courtyard has a principal room with 5 bays wide, flanked by wing rooms on both sides. On the south of the third courtyard is the drooping flowers gate composed by a pointed-topped facet and a coiling facet. On the north end is the posterior shielding storeyed building. The garden in the north of Prince Gong's Residence covers 2.6 hectares and is named Cuijin Garden. It is of high standing on account of its layout and distinct design. There are corridors, pavilions, artificial hills, ponds, trees and flowers distributed throughout the garden. There is an opera stage of great value on the south of eastern axis, which is the unique preserved one among Prince Residences in Beijing.

府邸前院
The Front Courtyard

府邸前院
The Front Courtyard

二宫门
The Second Gate

嘉乐堂
The Jiale Hall

府邸随墙门
The Gate on the Wall

嘉乐堂前廊
The Front Veranda of the Jiale Hall

戗檐砖雕（居家欢乐）
Brick Carvings on Qiangyan

戗檐、花盘子砖雕（梅鹿同春）
Brick Carvings on Qiangyan and
Huapanzi

多福轩西配殿
The Western Wing Hall of the
Duofu Hall

葆光室东配殿
The Eastern Wing Hall of the
Baoguang Hall

锡晋斋南立面
The Xijin Hall

后罩楼(宝约楼、瞻霁楼)
The Shielding Building

后罩楼为硬山调大脊，灰筒瓦屋面。楼前檐出廊，二层置荷叶净瓶木栏杆，其下为云纹木挂檐板，檐下饰以苏式彩画。此楼后檐墙二层开形状各异的什锦窗，砖雕窗框做工精美。楼下中间开过道门，可通向王府花园。

后罩楼东转角
The Eastern Turning Part of the Shielding Building

锡晋斋歇山式抱厦
**Bays with Gable-and-Hip Roofs of the Xijin Hall**

锡晋斋面阔七间，过垄脊灰筒瓦，前出廊，后出抱厦五间。正中三间大厅内三面设二层暗楼，楼内雕有华美细致的楠木碧纱橱、槛窗、栏杆等豪华装饰以及鼓墩式覆莲柱础，为和珅仿紫禁城宁寿宫而建，是他逾制获罪的重要证据。

锡晋斋东配殿（乐古斋）
**The Eastern Wing Hall of the Xijin Hall**

萃锦园大门
The Front Gate of the Cuijin Garden

萃锦园大门券肩及山花部分雕刻精美的
中式卷草图案。石刻楷体门额南面书"静
含太虚",北面书"秀挹恒春"。大门两侧
用青石堆叠假山,正中矗立一座高5米的
太湖石,上刻"独乐峰"三字。

安善堂
The An`shan Hall

韵花簃
Yunhuayi

水榭
The Pavilion at the Center of the Water Surface

沁秋亭
The Pavilion with Cup Flowing Through

沁秋亭内景
Interior of the Pavilion with Cup Flowing Through

邀月台敞厅
The Yaoyuetai Opening Hall

爬山廊
The Climbing Corridor

养云精舍，因外形似蝙蝠俗称"蝠厅"。
The Bat Hall

"蝠厅"翼角
The Roof Corner
of the Bat Hall

"蝠厅"彩绘
Coloured Paintings
of the Bat Hall

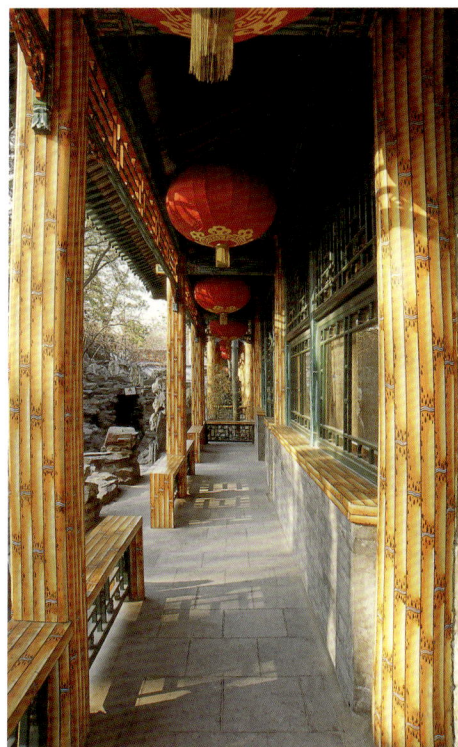

"蝠厅"廊外
The Veranda of the Bat Hall

大戏台内景
Interior of the Opera Theater

大戏楼侧面
The Side of the Opera Theater

# 醇亲王北府(醇亲王府)
# Prince Chun's North Residence

醇亲王北府位于北京市西城区后海北沿44号,又称摄政王府,末代皇帝溥仪便出生于此。2006年被公布为全国重点文物保护单位。

这里原是清康熙年间重臣明珠的府邸,后为清乾隆朝权臣和珅据为别院。嘉庆四年(1799年),嘉庆将其赐给兄长成亲王永瑆,成亲王府按王府规制,进行了大规模改建。后来,慈禧太后又将此府邸赐给的光绪皇帝的生父醇亲王奕譞,成为醇亲王北府。奕譞病故后其子载沣(光绪帝之五弟)袭爵。载沣之长子溥仪登基后,又成为摄政王府。

醇亲王府坐北朝南分为东、中、西三路,中路是其主体建筑。街门五间,进入外院后为府门,府门后为正殿,即银安殿;正殿后是一组自成院落的屋宇,属后寝部分;最后为后罩楼。中轴线建筑体量较高大,覆以绿琉璃瓦,显得雄伟庄严,但因占地宽敞,殿宇肃穆,不适宜生活居住,只于各种庆典时在此举行仪式,后寝成为供奉神佛和远祖的神殿。东路建筑很少,主要是家祠、佛堂及一些从属建筑。在其东墙外的又一院落则为王府马号所在(即今聋哑学校)。西路有两组院落并列,是面积较小的四合院落,适合日常起居,也是醇亲王府的活动中心。

王府西路花园,是保存最完好的王府花园之一。花园的庭院式建筑坐落于假山和湖水之间,以亭、轩、山、水闻名。主要建筑有益寿堂、畅襟斋、恩波亭等。

Located at No. 44 the northern bank of Houhai Lake in Xicheng District, the Prince Chun's Residence is also called the Prince Regent's Residence. It was listed as a national key relic under special preservation in 2006.

It was originally the residence of Mingzhu. During the reign of Emperor Qianlong, it was occupied by Heshen. In 1799, Emperor Jiaqing had Heshen executed and bestowed it to his brother Yongxing, known as Prince Cheng's Residence. It was rebuilt according the standard for a prince's residence. Then, Empress Dowager Cixi bestowed it to Yixuan, the father of Emperor Guangxu, to become Prince Chun's North Residence.

The Prince Chun's Residence, which faces south, are divided into the middle, the eastern, and the western axes. On the middle axis, from the south to the north, are the main buildings. The main hall flanked by wing rooms on the eastern and western sides is also called Yin'an Hall. On the end of the north is the posterior shielding storeyed building. Roofs of these imposing and magnificent buildings are covered with green glazed tiles. A variety of ceremonies were held here.

On the western axis is the garden, which is one of the best-preserved prince's gardens. Located among rockeries and lakes, it is famous for pavilions, hills and water.

府门
The Front Gate

银安殿鸱吻
Zhiwen at the End of the
Main Ridge of the Yin'an
Hall

银安殿
The Yin'an Hall

银安殿明间
The Yin'an Hall

银安殿角檐
The Roof-ridge of the Yin'an Hall

银安殿脊兽
Decorative Figurines of Immortals and
Mythical Beasts on the Eave Extensions of
Roof Cornef of the Yin'an Hall

寝门
The Sleeping Gate

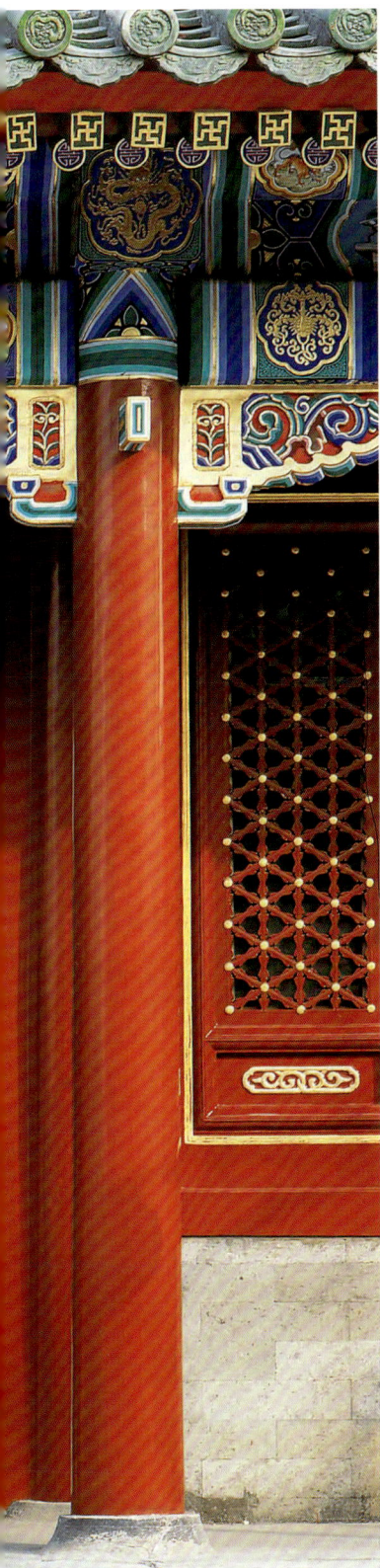

西侧翼楼
The Western Wing Storied Building

南楼
The Southern Storied Building

畅襟斋
The Changjin Hall

恩波亭
The En'bo Pavilion

恩波亭建在花园南湖东北的长廊上，为成亲王永瑆所建，是嘉庆皇帝破例恩准成亲王府花园引玉河水进园，永瑆建此亭以谢皇恩。亭内南檐上悬木匾，篆书"恩波亭"，这是因为皇宫在花园南面，建亭的主人要面南而谢。此亭两面临水，亭为单檐六角攒尖顶，亭南北两侧与长廊相接。

恩波亭
The En'bo Pavilion

# 醇亲王南府(醇王府)
# Prince Chun's South Residence

醇亲王南府位于北京市西城区太平湖东里。1989年被公布为西城区文物保护单位。

此府最初为乾隆第五子永琪的荣亲王府。咸丰九年(1859年)奕譞分府出宫，居此。同治十一年(1872年)奕譞晋醇亲王，府称醇亲王府，俗称七爷府。同治十三年(1874年)同治帝驾崩，奕譞的次子载湉嗣位，年号光绪。因光绪帝生在这里，是为"潜龙邸"。根据清制，皇帝的出生地应"升为宫殿"，因此光绪十四年(1888年)将什刹海后海北岸原棫贝子府赐予奕譞，因为此府在南，新府在北，故此府俗称南府。奕譞死后，太平湖的醇亲王府前半部改建为醇亲王祠，后半部仍作为"潜龙邸"。

府坐北朝南，分中路和东西二路及花园。现中路府门五间，两侧有八字影壁，内有东西两座阿斯门。中路还保存有二府门、寝殿、后楼以及花园部分建筑。二府门面阔五间，歇山顶黄琉璃筒瓦绿剪边屋面，是作为"潜龙邸"后改建的。寝殿五间，绿琉璃瓦硬山调大脊。后罩楼五间，是光绪皇帝的出生地。

Located at Taipinghu Dongli in Xicheng District, the Prince Chun's South Residence was listed as a Xicheng District's relic under preservation in 1989.

The residence was at first built as the residence of Yongqi, the 5th son of Emperor Qianlong, known as Prince Rong's Residence. In the 9th year of the reign of Emperor Xianfeng(1859A.D.), Yixuan moved out of the Forbidden City and lived here. In the 11th year of the reign of Emperor Tongzhi(1872 A.D.), Yixuan was conferred the title of Prince Chun, naming it as Prince Chun's Residence, also called the Seventh Prince's Residence. After Zaikuo, Emperor Guangxu, the second son of Yixuan ascended the throne, the place should be converted to the palace. Therefore, in 1888, Emperor Guangxu bestowed Beizi Su's Residence at the northern bank of Houhai Lake to Yixuan. To the south of the new residence, the residence was also known as South Residence.

Facing south, the residence is divided into the middle, the eastern and the western axes and the garden. On the middle axis, the gate is 3 bays wide, flanked by Ba Zi screen walls. Inside the gate, there are two gates leading to the eastern and western sides courtyards respectively. The principal structures, from the front to the rear, are the Yin'an Hall, the rear hall, the posterior shielding storeyed building, the garden and so on.

东阿斯门(即东南门)
The Southeastern Gate

二府门
The Second Gate

西跨院一进正房
The Principal Room in the
First Courtyard on the
Western Side Courtyard

# 庆王府
# Prince Qing's Residence

庆王府位于北京市西城区定阜大街3号。坐北向南，西临德胜门内大街，东接松树街，北界延年胡同，呈长方形。府内建筑多有改动，现中路只余后寝一座，东部亦已改建，只有西部建筑基本保存完整，昔日王府的规模和气派仍依稀可见。1984年被公布为北京市文物保护单位。

第一代庆亲王是乾隆帝的第十七子永璘，其王府本位于"三转桥，系和珅宅"。永璘去世后，根据清代爵位世袭递降的原则，其孙奕劻降至辅国将军爵，搬进原道光朝大学士琦善在定府大街的宅邸。奕劻善于逢迎，深得慈禧太后重用，光绪二十年（1894年）晋封为亲王。奕劻利用贪污的巨款，对庆王府大加修缮。在府内大兴土木，修建了绣楼和戏楼等处，建筑华丽精致。其中房屋分五个大院落，大小楼房约近千间。大门口是纯粹封建王朝的特殊形式，朱红大门。院内主房有九处，高大如宫殿，只是屋顶为灰瓦而不是琉璃瓦。

庆王府自东向西并列五套院落，奕劻住在西边的两套院子。院内有一座精美的二层绣楼至今仍保存十分完好，是西院的主要建筑之一。靠西墙的为后园，园内有一座二层戏楼，气派宏伟，可容纳三四百人之多，然而此楼不幸已被烧毁，现在其遗址上修建起一座礼堂。庆王府不仅房屋高大华丽，而且各个厅堂多悬匾额，如"宜春堂"、"承荫堂"、"乐有余堂"、"契兰斋"等。

Located at No.3 Dingfu Street in Xicheng District, the Prince Qing's Residence, which faces south, is in the shape of a rectangle. At its northern end lies the Yannian Hutong, its eastern end lies the Songshu Street, and its Western lies the Deshengmennei Street. From buildings on the western axis been partly preserved, we can still see the scale and style of the mansion, the site of which was listed as a Beijing's relic under preservation in 1984.

The first Prince Qing, named Yonglin, was the 17th son of Emperor Qianlong. His grandson Yikuang was not powerful at first, but was gradually appreciated by Empress Dowager Cixi. In the 20th year of the reign of Emperor Guangxu (1894 A.D.), Yikuang was conferred the title of Commandary Prince. He amassed large amount of taels of silver and enlarged the residence, which consisted of five courtyards from the east to the west, having about 1,000 buildings.

绣楼入口处
The Entrance of the Storied Building

王府绣楼
The Storied Building

# 孚王府
# Commandary Prince Fu's Residence

孚王府位于北京市东城区朝阳门内大街137号。2001年被公布为全国重点文物保护单位。

孚王府最初为康熙皇帝十三子允祥的怡亲王府。允祥病逝之后，原宅邸改建成贤良寺。而后雍正帝另赐第二代嗣王弘晓建新府于此。为与旧府相区别，世人习惯称弘晓府邸为怡亲王新府。咸丰帝即位（1851年），封道光皇帝第九子奕譓为孚郡王。同治三年（1864年），将原怡亲王府赐予奕譓，是为孚郡王府。同治十一年（1872年）奕譓晋封亲王，因其排行第九，故此府俗称"九爷府"。光绪三年（1877年），奕譓亡故。次年，以载澍为嗣，袭贝勒爵，故此府时称"澍贝勒府"。光绪二十三年（1897年），载澍因获罪被革去爵位，而后此府由奕譓之兄淳亲王奕誴之孙溥伒所有。

孚王府平面近似长方形，约6000平方米，大体上可分为中、西、东三路。中路为主要建筑所在，是王府的办公、会客和王爷起居的场所。自南而北建有大门、银安殿、后殿、寝殿，最后是后罩楼。西路建筑是由若干个四合院组成，院内建筑体量适中，院与院相对独立且又相互联系，平面布局完整，从使用上看，这组建筑应是王府眷属的居住区。东路建筑因毁坏较为严重，很难看出原有的格局。不过，从史书记载看，东路原属府库厨厩及执事房舍。

纵观整个王府，总体布局严谨规整，主次分明，虽遭受不同程度的破坏，尚保存其原建筑体系，尤其主体建筑保存完整。孚王府的建筑基本上可以说是个大型四合院。王府的多进院落，实际上就是一套四合院落的组合。门房正面是正殿，正殿两边是东西配殿，再往后又进入寝殿院落，东西两侧有跨院，最北面是后罩楼。其形制完全符合《大清会典》中规定的王府规制，是研究清代王府的典型实物资料。

Located at No.137 Chaoyangmennei Street in Dongcheng District, the Commandary Prince Fu's Residence was listed as a national key relic under special preservation in 2001.

The residence was primarily built as new residence of Prince Yi. After the death of Prince Yi, the 13th son Yunxiang of Emperor Kangxi, his residence was changed into Xianliang Temple. In 1864, Emperor Tongzhi bestowed it to Yihui, known as Prince Fu's Residence. In the 11th year of the reign of Emperor Tongzhi (1872 A. D.), Yihui was conferred the title of Prince Fu and as he was the 9th son in the family, the place was also called the "Ninth Prince's Residence". In 1878, Zaishu, the son of Yihui, inherited the rank of Beile, using it as his Beile residence and referring it as Beile Shu's Residence.

The residence shaped like a rectangle, covers an area of about 6,000 square meters. On the whole, it can be divided into the middle, the eastern and the western axes. On the middle axis, the principal structures, from the south to the north, are the gate, the Yin'an Hall, the rear hall, the inner quarters and the posterior shielding storeyed building.

内垣门（二府门）
The Second Gate

银安殿
The Yin'an Hall

翼楼
The Wing Storied Building

转角房
The Room in Turning Part

后殿
The Rear Hall

寝殿
The Sleeping Hall

寝殿
The Sleeping Hall

寝殿配殿
The Wing Hall of the Sleeping Hall

廊心墙砖雕(凤栖牡丹)
Brick Carvings on the Wall

# 郑王府
# Prince Zheng's Residence

郑王府位于北京市西城区西单北侧的大木仓胡同35号，是京城规模最大的王府之一。1984年被公布为北京市文物保护单位。

第一代郑亲王名济尔哈朗，是清初著名的"八大铁帽子王"之一。郑王府创建于清军进关之初。据说，该址原为明初功臣姚广孝府邸，后济尔哈朗因建府殿基逾制，又擅用铜狮、龟、鹤，遭弹劾后而罢官罚款。府邸建成后，历代袭王有所修缮或扩建，最重要的是第八代郑亲王德沛对花园的扩建，园名惠园，传是李渔的手笔，为清代著名园林。清末十三代郑亲王端华和怡亲王载垣及肃顺等同为"顾命八大臣"，"辛酉政变"后被赐自尽籍没家产。同治十年（1871年）又发还给已恢复世爵的庆至，复为郑亲王府。

郑王府坐北朝南，原布局分东、中、西三路，东部前躯突出，是主要殿宇所在，与其他王府以中路为主要建筑不同，王府中路原来有多组院落，为生活起居区，现在已经拆毁。西部花园是北京最大、最美的王府花园，可惜现已拆除建为二龙路中学。现仅东路残留有大门、二宫门（仪门）、正殿和乐堂、东西配楼和寝殿。府内原有后罩楼和一些附属建筑被拆除。

府门
The Front Gate

Located at No.35 Damucang Hutong to the north of Xidan in Xicheng District, the Prince Zheng's Residence was one of the largest mansions in Beijing and was listed as a Beijing's relic under preservation in 1984.

The first Prince Zheng, named Jirhalang, was one of the distinguished Eight-Iron-Hamlet-Princes in the early years of the Qing Dynasty. Built shortly after the Qing Dynasty was founded, the residence was renovated and expanded by later hereditary princes. The most important expansion of its garden was made by the 8th hereditary Prince Zheng, who renamed it as Huiyuan Garden which turned out to be the famous garden in the Qing Dynasty.

The residence, which faces south, was divided into the middle, the eastern and the western axes. On the eastern axis are the main buildings, which are different from other residences that the main buildings stand on the middle axis. The remaining buildings are the front gate, the ceremonial gate, the main hall and the rear hall.

石狮
The Stone Lion

丹陛石
Way Stair Slab

二宫门（仪门）
The Second Gate

垂脊走兽
Decorative Figurines of
Immortals and Mythical
Beasts on the Eave
Extensions of Roof Corner

山墙砖雕
Brick Carvings

彩绘
Coloured Paintings

彩绘
Coloured Paintings

院内游廊
Corridor in the Courtyard

垂花门
The Drooping Flowers
Gate

正寝殿(逸仙堂)
The Sleeping Hall

# 克勤郡王府
# Commandary Prince Keqin's Residence

克勤郡王府位于北京市西城区新文化街（原石驸马大街）西口路北53号，为清初"八大铁帽子王"（即世袭罔替）之一克勤郡王的府邸。1984年被公布为北京市文物保护单位。

克勤郡王府由努尔哈赤次子代善的孙子罗洛浑及重孙罗科铎修建。民国初年，曾售给了国务总理熊希龄。后熊希龄和夫人朱其慧将财产捐赠给北京救济会，至今府内尚存捐赠刻碑。建国后，克勤郡王府一直由学校使用。2001年11月开始对克勤郡王府进行了修复工程，使王府建筑基本恢复了旧日风貌。

该王府总建筑面积为3700多平方米。府外砖砌大影壁一座，是王府特权的标志物，只有王府能在街对面建如此高大的影壁，一般府邸只能建在门内或门两侧。府门五间三启门，一进院寝门三间，前出月台，汉白玉石护栏，东西各连看面墙。院内东西翼楼各五间，楼又各连庑房各七间，寝院内形成了开阔的小广场。二进院内正殿五间，为王府的寝殿，东西配殿各五间。第三进院为后罩房七间，东西厢房各五间，正殿与配殿间连转角房。

Located at No. 53 to the north of the western end of Xinwenhua Street (originally Shifuma Street) in Xicheng District, the Commandary Prince Keqin's Residence was the house of Commandary Prince Keqin, one of the Eight Iron-Hamlet Princes in the early years of the Qing Dynasty. It was listed as a Beijing's relic under preservation in 1984.

Built by Luoluohun and Luokeduo, the grandson and the great-grandson of Dai Shan (the 2nd son of Nurhachi) respectively, the residence was sold to Xiong Xiling, the State Premier in the early years of the Republic of China. Then, Xiong and his wife donated all their possessions to the Relief Association in Beijing. Now a stele about the donation still stands in the house. Since the foundation of the People's Republic of China, the residence have been used by schools.

The residence covers an area of more than 3,700 square meters. Located outside the front gate, the screen wall symbolizes the privilege of mansions for the reason that only mansions have screen walls at the opposite side of the gate. For ordinary residences, the screen walls could

府门
The Front Gate

only be built inside the front gates or at the eastern and western sides of the gates. The gate of the residence is 5 bays wide with 3 pairs of doors at the center. In the second courtyard, the main hall in 5 rooms, also called Yin'an Hall, on each side to the east and west flanked by wing rooms with 5 bays wide. In the third courtyard, the posterior shielding storeyed building flanked by wing rooms with 7 bays wide and on each side is 5 bays wide.

西翼楼及厖房
The Western Wing Storied
Building and the Hip Room

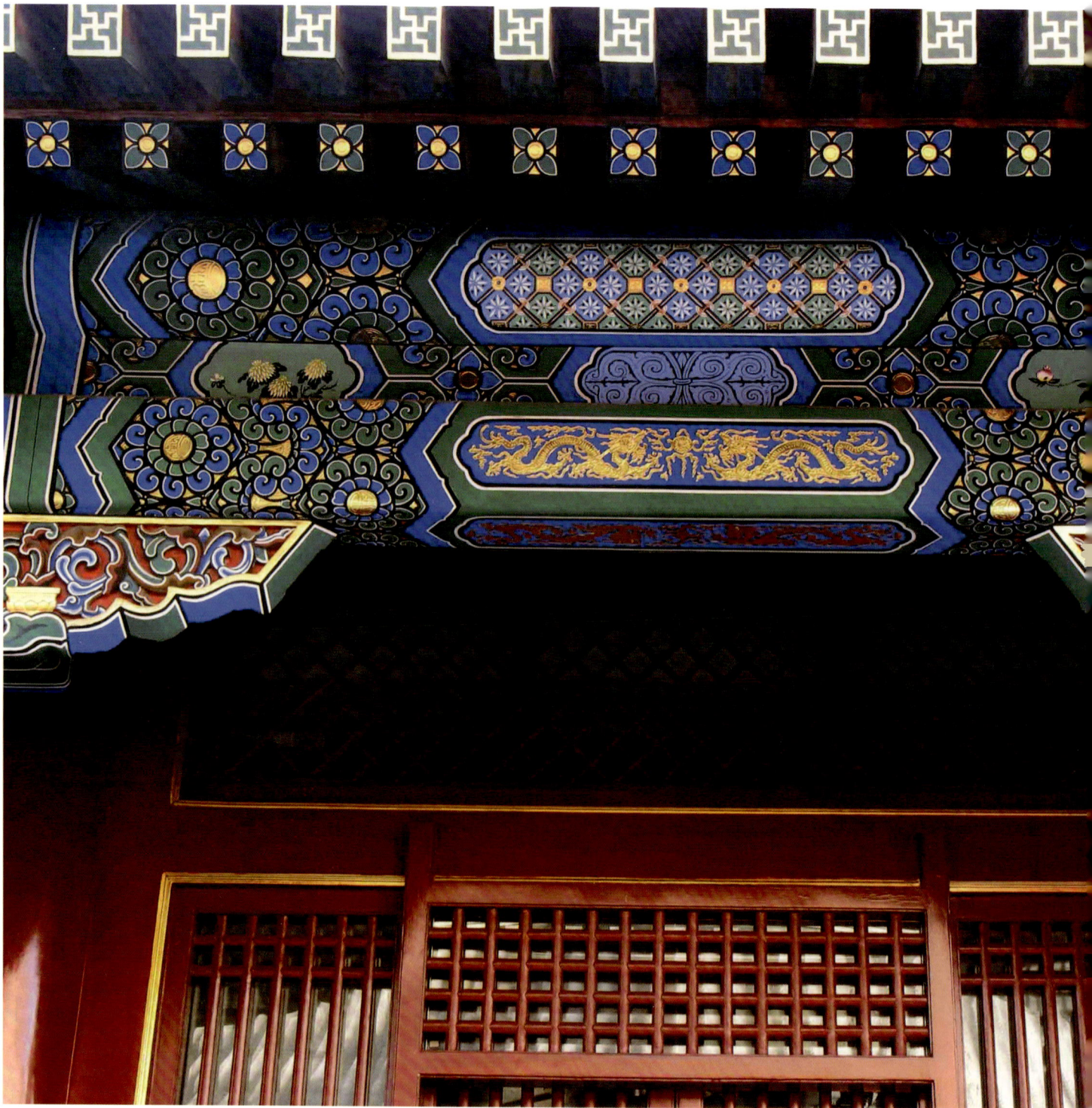

翼楼彩画
Coloured Paintings of the Wing
Storied Building

翼楼雀替、彩画
Carved Angle Braces and Coloured Paintings
of the Wing Storied Building

寝殿
The Sleeping Hall

寝门
The Sleeping Gate

寝殿东配殿
The Eastern Wing Hall of the
Sleeping Hall

后罩房
The Shielding Room

# 和敬公主府
# Princess Hejing's Residence

和敬公主府位于北京市东城区张自忠路7号，为乾隆第三女和敬公主下嫁后的赐邸，是北京保存下来的为数不多的公主府之一。1984年被公布为北京市文物保护单位。

该府原是王府建制，自外垣以内有正门、正殿、后寝、后楼和东西配房等附属建筑具备。以后爵位虽逐世降袭，府邸却保存下来，只是脊兽有所变动，将鸱吻改作望兽，以区别于王府。民国后该府成为北洋政府陆军部所在地，后寝部分进行了改扩建，但并未损及原有布局，主要建筑仍保存晚清风貌。

和敬公主府分为中、东、西三路。中路自外垣以内有府门三间，门外八字影壁，中心和四岔雕花卉。第一进院有过厅三间，过厅东西有配殿各三间，此院在公主府时期是传达室功能，府内的执事人员在此办事。

第二进院正殿五间，梁架绘制旋子彩画，前檐柱间装饰雀替，是公主府的正殿。院内东西配殿各五间，共同组成了公主府的外朝部分。

第三进院正殿五间，硬山顶筒瓦屋面，梁架也绘制旋子彩画，廊柱间也装饰雀替，是公主府的正寝殿，公主、驸马居此。院内东西配殿各五间，硬山顶筒瓦屋面，绘制等级较低的苏式彩画，前檐明间各带抱厦一间。

最后一进院为后罩楼七间，是公主府的绣楼，楼为两卷勾连搭形式，即一座房子为了加大空间再在后面加上一座形成"m"形屋顶，这种形式在绣楼中很少见。

西路保存基本尚好，院内游廊回环，东路已经变迁较大了。

府门
The Front Gate

Located at No. 7 Zhang Zizhong Street, this was the house conferred to Princess Hejing, the 3rd daughter of Emperor Qianlong, after she got married. It is one of the few preserved princess's residences and was listed as a Beijing's relic under preservation in 1984.

The house was originally built according to the standard for a prince's residence. The principal structures are the front gate, the main hall, the inner quarters, the posterior shielding storeyed building and the eastern and western wing rooms. Then, although the rank of prince degraded, the house still remains. Only Chiwen, the beast shape sculpture, on the ridge of the roof was replaced with Wangshou to make difference with the prince residences. The main buildings date back to the late Qing Dynasty.

The Princess Hejing's Residence is divided into the middle, the eastern, and the western axes. On the middle axis, the BaZi screen wall carved with flowers is located outside the front gate with 3 bays wide.

In the second courtyard stands the main hall with 5 bays wide, flanked by wing rooms on the eastern and western sides. The beams inside and outside the hall are decorated with polychrome paintings. All these compose the outer court.

In the third courtyard stands the main hall with 5 bays wide, flanked by wing rooms on the east and west sides. The flush gable roof is covered with tube-shaped tiles. The princess and her husband once lived here.

In the last courtyard stands the posterior shielding storeyed building with 7 bays wide.

On the western axis, the well-preserved courtyard is surrounded with corridors.

后罩楼
The Shielding Building

# 涛贝勒府
# Beile Tao's Residence

　　涛贝勒府位于北京市西城区柳荫街25、27、乙27号，是北京保存较为完好的贝勒府。1995年被公布为北京市文物保护单位。

　　涛贝勒府原系清康熙第十五子愉郡王允禑府邸。光绪二十八年(1902年)，醇贤亲王奕譞第七子载涛过继给钟郡王奕诒为嗣，承袭贝勒爵，迁居于此，称作涛贝勒府。涛贝勒府地处恭王府之西，庆王府之东，辅仁大学之北，是什刹海保护区内重要的文物建筑。现涛贝勒府中路、东路仍保持原格局，西路建筑无存，花园形制尚存。

　　该府坐北朝南，布局分东、中、西三路，西路已毁，东路保存五进院落。第一进南房三间，两侧有倒座各五间。东西两侧有厢房三间。北侧有正房五间，檐下置单昂三踩斗拱，前出廊，柱间带雀替。二进院东西两侧有厢房五间。北侧有正房五间。第三进院有正房五间，正房前有三出陛的月台一座，正房左右耳房各三间。东西厢房各三间。第四进院正房五间，东西厢房各三间。第五进院有排房共十三间，东侧七间，西侧六间。中路现存三进院落。第一进院南房七间，歇山顶筒瓦屋面，四周回廊，东西厢房各三间。北侧正房三间，两侧耳房各三间。第二进院有正房七间，左右耳房各二间。东西厢房各三间，均为双卷勾连搭形式。第三进院后罩房九间。

Located at No.25.27 Liuyin Street in Xicheng District, the Beile Tao's Residence is a relatively best-preserved Beile's residence in Beijing. It was listed as a Beijing's relic under preservation in 1995.

**It was originally the residence of Yunyu, Commandary**

东路一进院正房
The Principal Room in the First
Courtyard on the Eastern Axis

府门
The Front Gate

Prince Yu, the 15th son of Emperor Kangxi of the Qing Dynasty. In the 28th year of the reign of Emperor Guangxu, Yixuan (1902 A.D.), Prince Chunxian had his 7th son Zaitao adopted by Yihe, Commandary Prince Zhong, inheriting the rank of Beile. The adopted son then moved to the Commandary Prince Yu's Residence, using it as his Beile residence.

The residence, which faces south, is divided into the middle, the eastern, and the western axes. The buildings on the western axis had been destroyed. There are five courtyards extant on the eastern axis one behind another. In the first courtyard, the principal room is on the south with 3 bays wide, to either side of which reversibly-set rooms with 5 bays wide is attached. On the eastern and western sides, there are wing rooms with 3 bays wide respectively. In the second courtyard, there is the room on the south with 5 bays wide. On the east and west sides, there are wing rooms with 5 bays wide respectively. In the third courtyard, the side rooms with 3 bays wide are attached to either side of the principal room with 5 bays wide. In the fourth courtyard, there is the principal room with 5 bays wide, flanked by wing rooms with 3 bays wide on both sides. On the middle axis, there are three courtyards extant one behind another. In the first courtyard, the room on the south is 3 bays wide with a hip roof carved with tube-shaped tiles. On the eastern and western sides are the wing rooms with 3 bays wide respectively. In the second courtyard, the side rooms with 2 bays wide are attached to either side of the principal room with 7 bays wide.

西部院落(迁建)
The Western Side Courtyard

东路三进院正房
The Principal Room in the
Third Courtyard on the
Eastern Axis

中路敞轩
The Opening Hall on the
Middle Axis

中路二进正房
The Principal Room in the
Second Courtyard on the
Middle Axis

# 那王府
# Prince Na's Residence

　　国祥胡同2号四合院，东至宝钞胡同，南依国兴胡同，是蒙古喀尔喀赛因诺颜部札萨克和硕亲王那彦图府邸的一部分。1984年被公布为北京市文物保护单位。

　　那彦图（？－1938年）系清康乾时期名将超勇亲王策凌的七世孙，同治十三年（1874年）袭亲王爵，是末代札萨克亲王，故该府俗称"那王府"。

　　现国祥胡同甲2号两个院落，是原王府中路最北边的两个院子，整个院落坐北朝南，东西并连，大门朝北。东院南端有一殿一卷式垂花门，悬山顶过垄脊筒瓦屋面。北侧正房面阔五间，硬山过垄脊灰筒瓦屋面，前出廊，两侧各有耳房一间。东西厢房各三间，其中西厢房为两卷勾连搭的过厅，与西院相通。院中四面带游廊，梁枋绘苏式彩画。院中有两座太湖石，高约1.85米。北侧有七间后罩房。

　　西院南侧为花厅，面阔三间带周围廊，歇山顶过垄脊灰筒瓦屋面。北房五间，双卷勾连搭结构，硬山过垄脊灰筒瓦屋面，前出卷棚顶抱厦三间。房内明间原有"退洗斋"匾额，两侧有落地罩、碧纱橱。东西耳房各二间。西厢房三间，前出廊，过垄脊筒瓦屋面。东厢房即与东院共用的过厅。

槛墙及垂花门
The Drooping Flowers Gate and Walls on Both Sides

西院正房
The Principal Room on the Western Side Courtyard

322

Located at No.2 Guoxiang Hutong, the Prince Na's Residence was the house of Nayantu, Prince Heshuo. It was listed as a Beijing's relic under preservation in 1984.

Two courtyards at No.2 Guoxiang Hutong, which face south, were on the north end of the mansion. On the south of the eastern side courtyard is the drooping flowers gate composed by a pointed-topped facet and a coiling facet with an overhanging gable roof covered with tube-shaped tiles and Guolong ridge. The principal room is 5 bays wide with a flush gable roof and Guolong ridge. The roof is covered with tube-shaped tiles, on either side of which reversibly-set rooms with 1 bay wide are attached respectively. On both sides are the wing rooms with 3 bays wide respectively. All rooms are linked through short-cut corridors with beams decorated with polychrome paintings. On the north is the posterior shielding room with 7 bays wide. On the western side courtyard, the principal room is 5 bays wide. The interior fittings include exquisite bed-like compartments and shieldings reaching the ground.

西院花厅
The Flowery Hall on the Western
Side Courtyard

太湖石
Rocks from Taihu Lake

# 霭公府
# Mr. Yu's Residence

霭公府位于北京市西城区西绒线胡同51号，又称绵贝子府，是现存较完整的贝子级府邸。1989年被公布为西城区文物保护单位。

霭公即溥霭，是清圣祖康熙皇帝最小的儿子诚亲王允祕的六世孙，是贝子绵勋之曾孙。诚亲王府在安定门内宽街(即今北京中医医院的位置)，由于其爵位是世袭递降，袭封至贝子绵勋之后，同治八年(1869年)王府被收回转赐给咸丰皇帝唯一的亲生女儿荣安固伦公主，作为其下嫁的府邸。绵勋便迁到了西绒线胡同，所以此府是贝子府的建制。至光绪二十八年(1902年)溥霭袭镇国公爵，此府改称霭公府。辛亥革命后售与著名爱国人士、银行家周作民(1884－1955年)作为寓所。1950年中华人民共和国监察部曾在此办公。1959年在此开办四川饭店至今。

此府坐北朝南，现有建筑面积近2000平方米。府分中、东、西三路。中路是主要建筑所在，由五进院落组成。在中路的南北纵轴线上开府门，前后有两座垂花门，作为内宅的宅门，显示主人的社会经济地位，又象征吉祥平安。东、西两路为附属用房。

The Mr. Yu's Residence is located at No.51 Xirongxian Hutong, also known as Beizi Mian's Residence. It was listed as a Xicheng District's relic under preservation in 1989.

Mr. Yu's, Puyu, is the great-grandson of Beizi Mianxun. Facing south, the residence covers an area of nearly 2,000 square meters. It is divided into the middle, the eastern and the western axes. On the middle axis are the main buildings which comprise five courtyards one behind another.

垂花门看正房
Looking at the Principal
Room from the
Drooping Flowers Gate

内院
The Inner Courtyard

内院
The Inner Courtyard

夜景
Night View

# 东城区张自忠路23号(孙中山行馆)

## The Quadrangle at No. 23 Zhang Zizhong Street in Dongcheng District

行馆位于北京市东城区张自忠路23号，建于清代后期。2006年被公布为全国重点文物保护单位。

此宅在明代位于铁狮子胡同内，是崇祯皇帝宠妃之父左都督田弘遇住宅的一部分。名歌妓陈圆圆曾在此歌舞，这里曾有她的梳妆台和卧室。田弘遇在此把陈圆圆赠给山海关总兵吴三桂。清康熙年间，田宅成为靖逆侯张勇的府邸，名"天春园"。道光末年，竹溪以万金买下天春园，然后大加修葺，将其更名为"增旧园"。其子曾写《增旧园记》，描写园中八景。宅邸范围南起铁狮子胡同，北至府学胡同，东距中剪子巷20余米，西迄麒麟碑胡同和交道口南大街，规模宏大。清末民初，豪宅随着主人家的败落也被逐步分割出售。

现该宅有三进院落，大门面阔三间，宅院内有正房、厢房和倒座房，四周环以游廊。建筑群完全按照中国传统建筑的格式，主要建筑都集中在南北中轴线上，附属建筑分列左右，建筑装修精美。花园位于建筑群的东北侧，园内建筑与园林中的植被有机地融合起来，花木扶疏，屋宇错落，游廊曲折，园中各景相映成趣，成为私家园林的特有风格。孙中山先生生前就住在花园前的正房里，房间分内外套间，有雕刻精美的落地花罩。

Built in the last years of the Qing Dynasty, the quadrangle was listed as a national key relic under special preservation in 2006.

In the Ming Dynasty, the residence was originally located at Tieshizi Hutong, as a part of the parents' home of Lady Tian, a high ranked imperial concubine of Emperor Chongzhen. The celebrated singer Chen Yuanyuan once used the place as her performing stage. During the reign of Emperor Kangxi, it served as the home of Zhang Yong, Duke of Suppressing Rebellions, called Heavenly Spring Garden. During the reign of Emperor Daoguang, Zhuxi bought the place with a huge sum of money and renovated the house and renamed Garden of Adding to the Old Scenery. His son once wrote The Records of the Garden of Adding to the Old Scenery, depicting eight scenes in the garden. The magnificent residence ranged from Tieshizi Hutong in the south to Fuxue Hutong in the north.

Now the residence consists of three courtyards one behind another. The main buildings are the gate, principal rooms, wing rooms and reversibly-set rooms all linked by corridors. According to traditional Chinese architectural style, the main building complex stands on the middle axis, flanked by additional buildings on both sides. The garden is located to the northeast of building complex.

牡丹厅
The Peony Hall

花园方亭
The Square Pavilion of the Garden

垂花门
The Drooping Flowers Gate

游廊
Corridor

# 宣武区珠市口西大街241号(纪晓岚故居)

# The Quadrangle at No.241 Zhushikou Street in Xuanwu District

此宅位于北京市宣武区珠市口西大街241号。2003年被公布为北京市文物保护单位。

该宅始建年代不详,原为大将军岳钟琪府邸,后为纪晓岚居所,其中著名的阅微草堂是纪晓岚的书房。2002年,政府投资对故居进行修缮,同年将此处辟为纪晓岚故居展览馆。

此宅坐北朝南,由两进四合院建筑组成。原有建筑群第一进有广亮大门、正房、倒座房、正房五间,后因修路等原因,部分建筑已拆除,但主体建筑保存完好。一进正房三间,中式构架,西式装修,硬山顶,合瓦屋面。南立面为民国初年修缮时改建的中西合璧形式装修。屋顶为镂空女儿墙,门窗为拱券形式,上雕精美图案。二进正房为阅微草堂,面阔三间,硬山顶,合瓦屋面,前出廊,隔扇门,支摘窗,木构架绘有彩绘图案。全院有游廊贯穿,主次分明,布局合理。前院保存有紫藤萝,后院种有海棠,均为纪晓岚亲手所植。

The quadrangle is a former residence of Ji Xiaolan, a grand secretary and renowned scholar during the reign of Emperor Qianlong, and was listed as a Beijing's relic under preservation in 2003.

The quadrangle was originally the home of a general named Yue Zhongqi and then it was Ji Xiaolan's home, the well known of which is a studying room called the Thatched Abode of Close Observations.

The house, which faces south, comprises two courtyards one behind another. In the first courtyard, there is the principal room with 3 bays wide, the door and windows of which shaped like arch are carved with exquisite patterns. In the second courtyard, the principal room called the Thatched Abode of Close Observations, is 3 bays wide with corridors, partition doors of rooms and windows which can be propped up and down. The wooden constructions have colored paintings. The whole

南房
The Reversibly-Set Room

quadrangle is connected with corridors. A wistaria in the first courtyard and a Chinese flowering crabapple tree in the second courtyard were planted by Ji Xiaolan himself.

紫藤
Wistaria

# 东城区圆恩寺后街7号四合院
# The Quadrangle at No. 7 Yuanensi Backstreet in Dongcheng District

该宅建于清代后期，原系清末辅国公载搜的府邸。1984年被公布为北京市文物保护单位。

载搜（1887－1933年）是乾隆皇帝十七子永璘之曾孙庆亲王奕劻的次子。载搜是诸兄弟中最好享乐的，他娶了一个叫红宝宝的名妓，为博得她的欢心，便按其意建造了这座既有中式四合院又有西式洋楼的庭院供其享受。

此院坐北朝南，由东、中、西三路构成。宅院中部是一座西洋式楼房，砖混结构，地下一层，地上二层半。前出门廊为主入口，由6根爱奥尼亚柱式支撑。楼体做过大规模的抗震加固，外貌有所改动。楼前有一圆形喷水池，池中堆砌假山，池旁点缀有部分圆明园石刻遗物。池东南侧有一混凝土结构的圆亭，8根陶立克柱承托半圆形穹顶。其东侧有一道贯通南北的假山为障墙，分割出东院。穿过山

洞，豁然开朗，由两进院落组成。第一进院有过厅三间，歇山顶，过垄脊，灰筒瓦，带周围廊。院子东、南面有游廊相连接，廊子东南角有一六角攒尖顶小亭。亭前堆有假山，周围遍植树木花草。第二进院有北房五间，硬山过垄脊，灰筒瓦屋面，前后廊。周围环以游廊。东侧游廊前有勾连搭形式的敞轩三间。西跨院为一座两进四合院。广亮大门一间，一进院倒座房西侧五间，东侧四间，清水脊合瓦屋面。东路北房三间，悬山顶勾连搭形式。西路过一殿一卷式垂花门进入第二进院。院内有北房三间，耳房西侧二间，东侧三间，东西厢房各三间，南侧各带厢耳房二间，灰筒瓦屋面，四周有游廊相连接。东厢房北侧开随墙门一座，门前有八字影壁。各房均为硬山过垄脊。

花园廊亭
The Pavilion in the Garden

Built during the last years of the Qing Dynasty, it was originally the house of Zaibo,a minister at the end of the Qing Dynasty and was listed as a Beijing's relic under preservation in 1984.

Built by Zaibo, the great-grandson of Emperor Qianlong, the residence is composed of the Chinese quadrangle and the western-styled storied building.

The residence, which faces south, is divided into the middle, the eastern and the western axes. On the central section of the residence stands the western-styled storied building built of bricks and concretes. In front of the building lies a round fountain, to the southeast of which lies a round pavilion built of concretes. The eastern side courtyard comprises two courtyards one behind another.

In the first courtyard, the principal room is 3 bays wide with a gable-and-hip roof and Guolong ridge. The roof is covered with grey tube-shaped tiles. In the second courtyard, the principal room is 5 bays wide with a flush gable roof and Guolong ridge, each with verandas in the front and the back, surrounded with corridors. The western side courtyard comprises two courtyards one behind another. Guangliang Gate is one bay wide flanked by reversibly-set rooms on both sides. On the eastern axis, the principal room is 3 bays wide with an overhanging gable roof. On the western axis, the drooping flowers gate composed by a pointed-topped facet and a coiling facet stands on the south of the second courtyard.

西跨院正房
The Principal Room on the Western Side Courtyard

正房侧面
The Side of the Principal Room

正房明间
The Major Room of the Principal Room

游廊
Corridor

爬山廊
The Climbing Corridor

# 西城区富国街3号四合院
# The Quadrangle at No.3 Fuguo Street in Xicheng District

西城区富国街3号四合院位于北京市西城区平安大街平安里路口南侧，原为清初名将祖大寿的宅邸。1995年被公布为北京市文物保护单位。

祖大寿(? －1656年)，辽东人，始为明将，后降清，因作战英勇，颇受清太宗的器重，亲授总兵官，隶汉军正黄旗。清兵入关后，祖大寿住进西城大桥胡同(即今富国街，后改祖家街)。此处原为祖大寿宅，后改建为祖家祠堂，清雍正八年(1730年)在此设八旗官学、正黄旗官学，乾隆三十四年(1769年)重修。

该宅现保存基本完整，院坐北朝南，为三进院落，占地2300平方米。大门为三启一形式，门外有代表官宦宅邸的石狮和上马石各一对。门内北侧有过厅七间。二进院有正厅五间，东、西配房各三间。三进院前有一垂花门，下置滚墩石一对，雕刻精美，气派非凡，后殿五间，东西耳房各二间，东西配房各三间。该院建筑均为过垄脊筒瓦大式硬山房，梁枋绘有旋子彩画，建筑质量很高，是清代官僚住宅的典型代表。

Located to the south of the intersection of Pinganli Street and Ping'an Street in Xicheng District, the quadrangle was originally the house of Zu Dashou, a renowned general in the early Qing Dynasty. Zu born in Liaodong, was at first the Ming general, later surrendered to the Qing Dynasty. Shortly after the Qing Dynasty was founded, Zu moved here. The house was rebuilt as the ancestral temple of Zu's family afterwards. Renovated in the 34th year of the reign of Emperor Qianlong (1769 A. D.), it was listed as a Beijing's relic under preservation in 1995.

Facing south, the quadrangle covers an area of 2,300 square meters and comprises three courtyards one behind another. The gate is flanked by two stone lions and two stepping stones for mounting or dismounting from a horse. On the north of the first courtyard stands the principal room with 7 bays wide. In the second courtyard, the principal room is 5 bays wide, flanked by wing rooms on

第三进院
The Third Courtyard

the eastern and western sides. In the front of the third courtyard stands the drooping flowers gate flanked by a pair of gate piers. The rear hall is 5 bays wide, to either side of which reversibly-set rooms with 2 bays wide is attached. On the eastern and western sides, there are wing rooms with 3 bays wide respectively. The building arrangement is typical of the residences for officials during the Qing Dynasty.

二进正房
The Principal Room in the Second Courtyard

三进门内侧
The Inner Side of the Third Gate

三进院垂花门
The Drooping Flowers Gate in the Third Courtyard

# 东城区帽儿胡同7号、9号、11号(可园)
## The Quadrangle at No.7, No.9 and No. 11 Maoer Hutong in Dongcheng District

位于北京市东城区鼓楼之南的帽儿胡同7号、9号、11号，是清末大学士文煜宅邸的一部分(还应包括13号)。2001年被公布为全国重点文物保护单位。

全部建筑分为东、中、西三部分，是一座集建筑和园林为一体的建筑群。其中7号、9号便是旧京颇具盛名的私家园林——"可园"，始建时仿苏州拙政园和狮子林。其造园思想和手法非常巧妙，亭桥阁榭、花木竹石、布置精巧，错落有致，具有极高的艺术价值。11号为文煜宅邸的主要院落。

从整体布局看，此院东路和中路以园林为主，西路以建筑为主，各建筑群体相互独立而又联系紧密，遥相呼应，相得益彰。主要建筑均为大式硬山合瓦顶、带排山沟滴，屋宇高大，庭院宽敞，几进院落布局严谨，又不失灵活，空间变化自然丰富。居住建筑与园林水石相间，表现了北京四合院庄重大方又稳中求变的建筑特色。从原有规模和保存现状看，文煜旧宅堪称北京现存官僚宅邸建筑中最具代表性的实例，具有极高的历史价值、文化价值。

帽儿胡同9号第一进院
The First Courtyard at No.9 Maoer Hutong

Located to the south of Drum Tower in Dongcheng District, the quadrangle was a part of the home of Wen Yu, a grand secretary at the end of the Qing Dynasty. It was listed as a national key relic under special preservation in June 2001.

With its gardens and buildings, the whole building complex is divided into the middle, the eastern and the western axes. The place at No.7&9 named Keyuan Garden was a famous private garden in the Qing Dynasty. With very skillful thoughts and techniques, buildings, plants and rocks are elaborately arranged here. The main courtyards of Wen Yu's house are at No. 11.

From the whole layout, on the middle and eastern axes is the garden; on the western axis are the buildings. The building complex is either independent or compact. The main buildings all have gable style roofs with solid tiles and saddle ridges. With high buildings and wide courtyards, the layout of whole quadrangle is not only rigorous but also flexible and full of spatial changes naturally, thus, it is the most typical residence among bureaucratic government officials preserved in Beijing.

前院敞轩
The Opening Hall in the Front Courtyard

后轩
The Rear Hall

前院方亭
The Square Pavilion in the Front Courtyard

后院敞轩
The Opening Hall in the Backyard

中厅
The Middle Hall

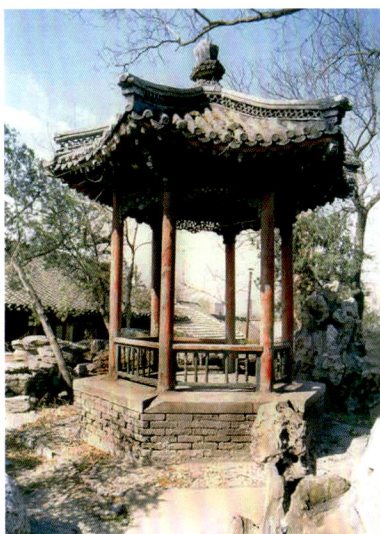

可园凉亭
The Pavilion in Keyuan Garden

假山
Rockwork

后院廊亭
The Pavilion in the Backyard

343

# 东城区府学胡同36号、交道口大街136号四合院

## The Quadrangles at No.36 Fuxue Hutong and No.136 Jiaodaokou Street in Dongcheng District

此宅是保存至今较为完整的大型四合院，由多组院落和花园组成，为北京地区四合院建筑的典型代表之一。1984年被公布为北京市文物保护单位。

明末此宅为崇祯帝田贵妃的娘家。康熙年间为靖逆侯张勇所有。清代末年，成为兵部尚书志和的府邸，民国时期同治帝的两位遗孀敬懿、荣惠皇太妃曾居于此，建筑有所改建。西宅曾为北洋军阀海军总长刘冠雄的官邸。此后为天主教神学院所有。

府学胡同36号原宅门三间，现已封堵，门外左右列上马石。门内为中路第一进院，现存东房、北房、南房及一幢面阔七间的二层仿古建筑。两侧与游廊相接。

内院可分为坐北朝南的三组院落。

中路第二进院北房为过厅。两侧原有配廊，以通东西两院。中路第三进院有垂花门一座，院内有北房、东西厢房。各房之间均有游廊相互连接。后有五间后罩房。

西路第一进有北房、两侧耳房，南房及西侧耳房。第二进院有东、西厢房。二、三进院之间有一垂花门。垂花门内有北房、东西耳房、东西厢房，各房之间均有游廊相连。第四进院为后罩房。

东院第一进院四周游廊，现南段已拆除，其余部分保存尚好，院内有敞厅一座，两侧开门与游廊相接。北侧有一垂花门连接二院。第二进院有东房、北房。第三进院有四间北房。

西宅即交道口大街136号，坐北朝南，原有门在麒麟碑胡同路北，大门封闭后，又于宅院西南隅辟门，临交道口南大街，此门为现北京市东四妇产医院的大门。此宅原有四进院落，今仅存三进。原第一进院现已改造，作为医院大厅。

第二至四进院呈长方形，从西侧走廊进入。第二、三进院均有北房、东西厢房。第四进院北房两侧有耳房、两侧有厢房，西房为二层建筑。院内环以游廊。

The quadrangles were the house of Zhihe, a minister of national defence at the end of Qing Dynasty. They were listed as a Beijng's relic under preservation in 1984.

In the end of Ming Dynasty, the house was the parents' home of Lady Tian, a high ranked imperial concubine of Emperor Chongzhen. During the reign of Emperor Kangxi, it was the home of Zhang Yong, Duke of Suppressing Rebellions. During the Republic of China, former Empress Jingyi and Empress Ronghui once lived here.

The quadrangle at No.36 Fuxue Hutong, which faces

宅门
The Front Gate

south, is divided into the middle, the eastern and western axes. The gate was 3 bays wide flanked by a pair of stepping stones for mounting or dismounting from a horse.

On the middle axis, the first courtyard has a principal room flanked by side rooms and a reversibly-set room. The second courtyard has eastern and western wing rooms. The drooping flowers gate stands between the second and the third courtyards. All the rooms are surrounded with short-cut corridors.

On the eastern axis, the drooping flowers gate is located between the first and the second courtyards. The second courtyard has a principal room and eastern wing room. In the third courtyard, the principal room is 4 bays wide.

The quadrangle at No.136 Jiaodaokou Street, which faces south, comprises three courtyards one behind another. This is a well-preserved quadrangle comprising several groups of courtyards and gardens. It is a typical quadrangle building in Beijing area.

上马石
Stepping Stone for Mounting or
Dismounting from a Horse

第一进院
The Front Courtyard

过厅
The Hall

游廊什锦窗
Windows of Various Forms

游廊
Corridor

垂花门(局部)
Section the of Drooping Flowers Gate

中路垂花门
The Drooping Flowers Gate on
the Middle Axis

花园
The Garden

# 东城区礼士胡同129号四合院
# The Quadrangle at No.129 Lishi Hutong in Dongcheng District

此宅位于北京市东城区灯市口附近的礼士胡同，房屋高大，颇为壮丽。1984年被公布为北京市文物保护单位。

这所宅子曾被误传为清乾隆年间大学士刘墉的宅邸。其实此为清末武昌知府宾俊的住宅，后几经易手，卖给天津盐商李颂臣，他交由朱启钤的学生重新设计，改建成如今之规模。

这所宅院坐北朝南，由住宅和花园两部分组成，占地面积约1200平方米。此宅虽为民国时期改建，但布局紧凑，设计精妙，墙体皆为磨砖对缝，建筑质量很高。

The splendid and magnificent quadrangle is located at Lishi Hutong near Dengshikou in Dongcheng District. It was listed as a Beijing's relic under preservation in 1984.

This house used to be mistaken as the residence of Liu Yong, a famous grand secretary during the reign of Emperor Qianlong. Actually, it was at first the residence of Bin Jun, a prefect of Wuchang at the end of the Qing Dynasty. It was sold many times and finally sold to Li Songchen, a salt businessman in Tianjin. Li asked Zhu Qiqian's student to redesign the quadrangle and changed it to its present scale.

The quadrangle, which faces south, covers an area of about 1,200 square meters and comprises the residence and the garden.

大门花墙
The Flowery Walls of the Front Gate

宅门
The Front Gate

一进院
The First Courtyard

游廊、什锦窗
Corridor and Windows of Various Forms

中路二进院
The Second Courtyard on the Middle Axis

二进院正房
The Principal Room in the Second Courtyard

廊心墙砖雕
The Brick Carving on Top of the
Opening at the End of the Corridor

花园水榭
The Pavilion at the Center of the Water Surface in the Garden

园中一景
A Scenic Spot in the Garden

# 西城区西四北大街六条23号四合院
# The Quadrangle at No.23 Xisi Beiliutiao Street in Xicheng District

该院为典型的中型四合院住宅，占地面积2500平方米，纵跨西四北六、七条，四进院落保存完整。1984年被公布为北京市文物保护单位。

该院建于清末民初，坐北朝南。门外有一字影壁，门前台阶两侧为一对上马石。广亮大门一间，两侧有倒座房共七间，东侧二间，西侧五间，清水脊合瓦屋面。门内有一坐山影壁。西侧有一殿一卷式垂花门一座，两侧连接看面墙，墙上嵌什锦灯窗，内侧为游廊。二进院有过厅五间，前后出廊，清水脊合瓦屋面。左右各有耳房二间，东西厢房各三间，南侧厢耳房一间，均为过垄脊合瓦屋面。

四周环以抄手游廊。北房明间隔扇门裙板雕刻《西游记》等古典小说的人物形象和花篮盆景图案，其东耳房一间为过道与后院相通，过道墙上布满"卍"字纹砖雕。第三进院正房五间，前后出廊，清水脊合瓦屋面。左右各有耳房二间。东西厢房各三间，南侧带厢耳房一间，前出廊，清水脊合瓦屋面。第四进院有后罩房九间，亦为清水脊合瓦屋面。三进院东侧有一跨院，院内有正房三间，前后出廊，清水脊合瓦屋面。东西厢房各三间，西耳房一间，均为过垄脊合瓦屋面。东侧有一栋二层楼，为后期改建。

垂花门及游廊
The Drooping Flowers Gate and Corridor

As a typical middle quadrangle, the quadrangle covers an area of 2,500 square meters. Composed of four courtyards one behind another, it was listed as a Beijing's relic under preservation in 1984.

Facing south the quadrangle was built in the last years of the Qing Dynasty and the early years of the Republic of China. Guangliang Gate, be 1 bay wide, is flanked by a pair of stepping stones for mounting or dismounting from a horse. Yi Zi screen wall is located outside the gate. Reversibly-set rooms are 7 bays wide with 5 bays to the west of the gate and 2 bays to the east of the gate. On the western side, there is the drooping flowers gate composed by a pointed-topped facet and a coiling facet linking separate walls on both sides. At the inner side of the walls are the corridors. In the second courtyard, the principal room is 5 bays wide with front and back corridors. The side rooms are 2 bays wide attached to both sides of the principal room. On the eastern and western sides, there are the wing rooms with 3 bays wide respectively. All the rooms are surrounded with short-cut corridors. The partition doors of rooms are carved with patterns of figures and potted landscapes. In the third courtyard, the principal room is 5 bays wide with front and back corridors. The side rooms are 2 bays wide attached to both sides of the principal room. On the eastern and western sides, there are the wing rooms with 3 bays wide respectively. In the fourth courtyard, the posterior shield room is 9 bays wide.

隔扇门
Partition Doors of Rooms

# 东城区东棉花胡同15号院及拱门砖雕
# The Quadrangle at No.15 Dongmianhua Hutong in Dongcheng District

该宅有一座砖雕拱门十分精美。2001年,15号院及其拱门砖雕被公布为北京市文物保护单位。

这里原是清末西安将军凤山的宅子,据传此人甚得慈禧宠信,且富于资财,故其宅也非常大,原东棉花胡同东半部,以及一条南北小横胡同几乎全是他的房产。宣统三年(1911年)凤山死后,其家也就此败落,房产被分割出售。

此院坐北朝南,原宅大门已拆除,现存广亮门是后开的。院内原一殿一卷式垂花门已改建成一间住房,二门即为此砖雕拱门。门为半圆拱形,高4米多,宽约2.5米,从金刚墙以上均为砖雕,上刻花卉及走兽。顶部为朝天栏杆,栏板上雕着松、竹、梅"岁寒三友"。拱券中间的汉白玉拱心石上雕刻"福到眼前"图案,笔法老练圆润。拱门外两侧雕有多宝阁,阁内雕有暗八仙等博古图案。整个拱门上的砖雕,布局严谨、凹凸得当,其做工之细,刀法之精,实属罕见。

这座拱门与其左右相连的民国式拱券窗平房建筑,配合紧密,浑然一体,拱门内为一四合院落,但因房屋年久失修,略显破旧。

Inside the quadrangle at No.15 Dongmianhua Hutong to the south of Jiaodaokou Street in Dongcheng District lies an arch gate with exquisite brick carvings. In 2001, the quadrangle at No.15 and the arch gate were listed as a Beijing's relic under preservation.

The quadrangle was originally the house of Fengshan, a general at the end of the Qing Dynasty. The arch gate, in 4 meters high and 2.5 meters wide, is decorated with brick carvings, the designs of which are composed of flowers and animals. The railing board on top of the gate is carved with pines, bamboos and plum blossoms. On either side of the gate is a picture of curio shelf carved with designs of 8 immortals presented in a concealed way.

院门砖雕(局部)
Section of Brick Carvings of the
Front Gate

院门砖雕
Brick Carvings of the Front Gate

院门砖雕（局部）
Section of Brick Carvings of
the Front Gate

院门砖雕（局部）
Section of Brick Carvings of
the Front Gate

# 西城区西四北三条11号四合院
# The Quadrangle at No.11 Xisi Beisantiao Street in Xicheng District

该院位于北京市西城区西四北三条东段，建于清末民初，建筑面积1800平方米，为花园式四合院，原为国民党政府委员、蒙藏委员会委员长马福祥（1876－1932年）的住宅。1984年被公布为北京市文物保护单位。

该院坐北朝南，贯通西四北三、四条，前后共五进院落，东侧还有一跨院为花园。广亮大门位于院落东南隅，合瓦清水脊屋面，倒座房六间，大门西侧五间，东侧一间。二进院正中有一殿一卷式垂花门，两侧连接游廊。二进院有正房三间，东西厢房各三间。三进院与第二进格局相同。第四进院有正房七间，东西耳房各二间，西侧有厢房三间。第五进院有后罩房十四间。第四进院有过道可通东侧跨院，为此宅的花园部分，有北房五间，西厢房六间，院内东北侧叠石为山，下有山洞，上置爬山游廊，连接东侧二层配楼，东配楼南侧有一八角形筒瓦攒尖顶小亭，立于假山之上。

Located to the eastern part of Xisi Beisantiao Street, the quadrangle was built in the last years of the Qing Dynasty and the early years of the Republic of China. Covering an area of 1,800 square meters, the quadrangle used to be the home of Ma fuxiang, a committeeman of Kuomintang Government. It was listed as a Beijing's relic under preservation in 1984.

Through Xisi Beisantiao and Xisi Beisitiao Streets, the quadrangle, which faces south, comprises five courtyards one behind another. The eastern side courtyard is a garden. Guangliang Gate is open at the southeastern corner of the quadrangle. The reversibly-set rooms are 6 bays wide with 5 bays to the west and 1 bay to the east of the gate. In the central section of the second courtyard stands the drooping flowers gate composed by a pointed-topped facet and a coiling facet, with corridors on both sides. The principal room is 3 bays wide. On the eastern and western sides, there are wing rooms with 3 bays wide respectively. In the fourth courtyard, the principal room is 7 bays wide with side rooms with 2 bays attached to either side. In the fifth courtyard, the posterior shielding room is 14 bays wide.

西厢房
The Western Wing Room

东跨院
The Eastern Side Courtyard

东配楼
The Eastern Wing Storied Building

# 西城区前公用胡同15号四合院
# The Quadrangle at No.15 Qiangongyong Hutong in Xicheng District

北京市西城区新街口前公用胡同15号四合院，原为清末内务府大臣崇厚(1826－1893年)的宅邸。1984年被公布为北京市文物保护单位。

该院共有东、中、西三组院落，除了中路的二进院落外，东西两路跨院各有三进院落，形式相似。中路府门三间，门两侧有雕刻精美的上马石一对。院内前部为花园，中间堆有叠石花坛。其北侧有花厅五间，为该院的主体建筑，是当时主人宴请宾客的场所，明间前出抱厦，可为戏台。此花厅的建造独具匠心，抱厦采用六檩卷棚顶，大气而不笨重，与南侧的花园融为一体，建筑四面开窗，两侧是精美的圆形花窗。廊檐下两侧置"松竹梅"砖雕，雕刻工艺精细，生动传神，保存较好。花厅抱厦两旁堆有假山石，两侧建有造型别致的月亮门。进入花厅东侧的月亮门便可来到后院。后院有正房三间，前后出廊，两侧耳房各二间，东西厢房各三间。院内四周环以游廊，建筑均为过垄脊合瓦屋面。

Located at No.15 Qiangongyong Hutong in Xicheng District, the quadrangle was originally the house of Chonghou, a minister at the end of the Qing Dynasty. It was listed as a Beijing's relic under preservation in 1984.

The quadrangle is divided into the middle, the eastern and the western axes. On the eastern and western axes, there are three courtyards one behind another respectively. On the middle axis, there are two courtyards one behind another. The gate is 3 bays wide, flanked by a pair of exquisite stepping stones for mounting or dismounting from a horse. The flower hall on the north of the courtyard is 5 bays wide and on its south serving as a small opera stage. Under the eaves of the flower hall are a set of beautiful brick carvings with patterns of pines, bamboos and plum blossoms. The flower hall is flanked with two unique moon-shaped gates. Entering the one on the east, you come to the second courtyard. The principal room is 3 bays wide each with verandas in the front and at the back, to either side of which side rooms with 2 bays wide are attached respectively.

西院垂花门
The Drooping Flowers Gate on the Western Side Courtyard

正房明间隔扇门
The Partition Doors of Rooms of the
Major Room of the Principal Room

游廊
Corridor

东跨院
The Eastern Side Courtyard

中院假山
The Rockwork on the
Middle Axis

# 东城区黑芝麻胡同13号四合院
# The Quadrangle at No.13 Heizhima Hutong
# in Dongcheng District

　　该宅东邻南锣鼓巷，南依沙井胡同，原为清光绪时四川总督奎俊的房产，是一座东西并连的大型四合院落。2003年被公布为北京市文物保护单位。

　　此宅坐北朝南，高台阶上有广亮大门一间，清水脊合瓦屋面，檐柱间带雀替。大门两侧有上马石一对，临街有一字影壁。门内亦有一字影壁一座，东西两侧为屏门，可通东、西路院落。西路一进院有倒座房八间半，清水脊合瓦屋面。北房为过厅八间，东侧第八间为过道。第二进院较小，正中有双卷勾连搭垂花门一座，两侧连接看面墙。三进院有正房三间，前后出廊。两侧有耳房各一间。东西厢房各三间，南侧有厢耳房各一间，建筑均为过垄脊合瓦屋面，四周有游廊连接各房。第四进院后罩房已翻建，院

内东、南侧尚存部分游廊。此院现从前鼓楼苑胡同辟门。东路院有倒座房十二间，北侧有一殿一卷式垂花门一座。二进院有正房三间，前后出廊，两侧有耳房各一间，东西厢房各三间，四周有游廊相连接，各房均为过垄脊合瓦屋面。第三进院有北房七间，清水脊合瓦屋面。东西各有平顶房一间。其东侧有二进跨院，有过垄脊合瓦北房各三间，为新改建。

　　此宅房屋的廊柱、檐柱粗大，已超过清《工部工程则例》的规定。院内建筑保存较好，格局完整，砖、石、木雕精美，很有特色。此宅院东侧原有大面积的花园，园中置有假山、游廊、亭轩、月牙河、珍贵树木等。

影壁
The Screen Wall

The quadrangle was originally the house of Kuijun, a satrap of Sichuan Province of the Qing Dynasty. At its eastern end lies the Nanluogu Alley. At its southern end lies the Shajing Hutong. It was listed as a Beijing's relic under preservation in 2003.

The quadrangle, which faces south, is divided into the eastern and the western axes. On the western axis, the buildings comprise four courtyards one behind another. On the eastern axis, the buildings comprise three courtyards one behind another. They all have been best-preserved.

明间隔扇门
The Partition Doors of Rooms of the Major Room

# 东城区鼓楼东大街255号四合院
# The Quadrangle at No.255 Guloudong Street in Dongcheng District

　　此宅为三进院落，做工考究，民国时期建造。2001年被公布为北京市文物保护单位。

　　此宅本为坐西朝东，大门位于草场胡同内。后改为坐北朝南形式。中间开广亮大门一间，两侧各有倒座房二间。一进院正房七间、东西厢房各三间，均为后期改建。二进院北面有一殿一卷式垂花门一座。垂花门两侧看面墙上有形态各异、雕刻精美的砖雕什锦窗。正对垂花门的南墙上，嵌有一座汉白玉石影壁，上刻有二龙戏珠图案。垂花门内为一方形大花园，花园当中有一座双层六边形汉白玉西式喷水池。池上每边角处均雕有一只小石狮，中央为一蟠龙柱式的出水口，顶端雕有四个龙头。花园北部正房面阔七间，后部出廊，前面五间吞廊。两侧各有耳房一间，均为硬山过垄脊，灰筒瓦屋面。室内装修异常精美，有镶嵌大理石的硬木隔扇，硬木雕花的落地罩、博古架。北墙明间砖雕凤凰、牡丹图案，十分精美，保存完整。后院西北角有后罩房四间。

　　该院落从布局上看应为原大型宅邸之附属花园，后来改建成了中西合璧形式。院内的砖雕、石刻以及室内的装饰均十分精美，保存之好，在现存北京四合院中实属罕见。

正房明间装修
The Decoration on the Major Room of the Principal Room

垂花门
The Drooping Flowers Gate

The quadrangle consists of three courtyards one behind another. Built during the Republic of China, it was listed as a Beijing's relic under preservation in 2001.

The quadrangle, which originally faced east, was changed into facing south. The Guangliang Gate, be 1 bay wide, is flanked by reversibly-set rooms with 2 bays wide on both sides. In the first courtyard, the principal room is 7 bays wide. On the eastern and western sides, there are the wing rooms with 3 bays wide respectively. To the north of the second courtyard stands a drooping flowers gate, through which you will come into a square great garden. In the garden lies a white marble western-style fountain which is in the shape of a hexagon with two tiers. The principal room is 7 bays wide. To the northwestern corner of the third courtyard stands the posterior shielding room with 4 bays wide.

**图书在版编目(CIP)数据**

北京古代建筑精粹/北京市文物局,
《北京古代建筑精粹》编委会编.—北京:
北京美术摄影出版社,2007.11
ISBN 978-7-80501-378-7

I.北… II.①北…②北… III.古建筑—简介—北京市
IV.K928.71

中国版本图书馆CIP数据核字(2007)第162371号

## 北京古代建筑精粹

BEIJING GUDAI JIANZHU JINGCUI

北 京 市 文 物 局 编
《北京古代建筑精粹》编委会

| | |
|---|---|
| 出　　版 | 北京出版社出版集团<br>北京美术摄影出版社 |
| 地　　址 | 中国·北京·北三环中路6号 |
| 邮　　编 | 100011 |
| 网　　址 | www.bph.com.cn |
| 发　　行 | 北京出版社出版集团 |
| 制　　版 | 北京顺诚彩色印刷有限公司 |
| 印　　装 | 北京印刷集团有限责任公司印刷一厂 |
| 版　　次 | 2007年11月第1版第1次印刷 |
| 开　　本 | 889×1194　1/16 |
| 印　　张 | 45 |
| 书　　号 | ISBN 978-7-80501-378-7/J·336 |
| 定　　价 | 980.00元(全两卷) |

质量投诉电话　010-58572393